W9-AHK-755

Troubleshooting Electronic Equipment Without Service Data

OTHER BOOKS BY ROBERT MIDDLETON:

New Ways to Use Test Meters: A Modern Guide to Electronic Servicing
New Digital Troubleshooting Techniques

Troubleshooting Electronic Equipment Without Service Data

Robert G. Middleton

Prentice-Hall, Inc.
Englewood Cliffs, New Jersey

Business and Professional Division

Prentice-Hall International, Inc., *London*
Prentice-Hall of Australia, Pty. Ltd., *Sydney*
Prentice-Hall Canada, Inc., *Toronto*
Prentice-Hall of India Private Ltd., *New Delhi*
Prentice-Hall of Japan, Inc., *Tokyo*
Prentice-Hall of Southeast Asia Pte. Ltd., *Singapore*
Whitehall Books, Ltd., *Wellington, New Zealand*
Editora Prentice-Hall do Brasil Ltda., *Rio de Janeiro*

Third Printing. . . . . . .November 1985

Library of Congress Cataloging in Publication Data
Middleton, Robert Gordon
Troubleshooting electronic equipment without service data.
Includes index.
1. Electonic apparatus and appliances—Maintenance and repair. I. Title.
TK7870.2.M535 1984 621.381'028'8 83-13772
ISBN 0-13-931097-5

Editor: George E. Parker

Printed in the United States of America

A WORD FROM THE AUTHOR ON THE UNIQUE, PRACTICAL VALUE THIS BOOK OFFERS

Meeting the challenge of troubleshooting audio, radio, television, and related electronic equipment without service data is the subject of this book.

Experienced electronic troubleshooters know that service data may not be available for various reasons:

1. The manufacturer sometimes releases a new product with the intention of following up with service data when the factory is less rushed.
2. It may be difficult to run down the address of a foreign consumer electronics product manufacturer.
3. There is always some lag time involved in the availability of service data from professional publishers.
4. Service data may never be issued for various proprietary brands of electronic equipment.
5. Numerous design changes are sometimes made for a particular production run, which make the available service data misleading.
6. Although service data may be available, the troubleshooter may decide that it doesn't pay to invest ten dollars for a small-ticket job.
7. Sometimes the available service data is so sketchy that it is of no real assistance in the troubleshooting procedure.
8. After a unit of electronic equipment has been serviced numerous times in various shops, the troubleshooter may be confronted with one or more nonstandard repairs with which he must contend, or he must return the unit to his customer on a concession loss basis.
9. Manufacturers "goof" once in a while and issue service data with technical errors, drafting errors, or incorrect entries. The professional troubleshooter needs to have the ability to spot such errors.

10. Service-data writers sometimes assume that the troubleshooter has sophisticated (and expensive) test equipment on hand, which in fact he does not. In turn, the troubleshooter is thrown upon his own resources, although service data are "available."

Many never-before-published troubleshooting techniques are explained in this book, with case histories to show how the new techniques are applied and evaluated in actual practice, step by step. Included among these unique troubleshooting techniques are:

high-speed differential temperature checker
"blind" troubleshooting procedures
self-oscillatory quick check
audio impedance checker
color code mapping
checkout of stage sequence
check of Thevenin impedance
measurement of Thevenin impedance in a low-level stage
capture effect
impedance check at battery-clip terminals
resonance probe
high-impedance tuned signal-tracing probe
identification of FM transformers in FM/AM receiver
heterodyne signal-tracing probe
addition or subtraction of DC voltages
converter voltages in oscillatory and nonoscillatory states
transceiver quick checks
measurement of quench frequency
check of incidental frequency modulation
identifying a "stone-dead" stage
impedance quick checks in sync circuitry

Troubleshooting without service data is both an art and a science. This is just another way of saying that the troubleshooter's mental attitude is as important as his or her technical know-how. Technicians who are comparatively experienced in conventional troubleshooting procedures sometimes have a mental block when confronted by a strange unit of electronic equipment—without any service data. This "mental block" occurs because the *fundamental approach* to trouble symptom analysis and testing procedure is different when no service data are at hand.

Most service shops keep a library of service data, and the general rule for technicians is to read the manual. However, many old-timers seldom do this. Over the years, they have come to know that all electronic equipment of a certain type has to work in a certain way, and that there are specific trouble symptoms for each section of the circuitry.

Furthermore, each manufacturer has his own particular style, so there is a family resemblance between his products. It is therefore quite feasible for the experienced troubleshooter to operate without service data, using methods similar to those described herein. This is not to say that service data should be ignored when available, but that the experienced technician can generally manage without it.

In this book, various tricks of the trade are described. Most of these "tricks" require both a thorough knowledge of the way in which circuits operate and sufficient experience to recognize when test-equipment readings are "in the ballpark." It is in the service shop that similar equipment in working order will most likely be available for comparison tests. In the home, comparison tests will generally be confined to stereo systems; in turn, more generalized approaches must often be followed.

The same observation applies to some test equipment, so that this book has somewhat more extensive application for service technicians than for hobbyists and experimenters. On the other hand, this is not to imply that this book is not for the hobbyist or experimenter—quite to the contrary, many of the test procedures and homemade test equipment units featured in the text will be of interest to everyone who is concerned with electronic circuitry.

Various facts that are taken for granted in conventional troubleshooting procedures may not be immediately apparent when no service data is at hand. For example, in the absence of service data, the technician cannot refer to a layout diagram and immediately point to an RF section, to an IF section, to a video section, to a sync section, to an audio section, and so on. He cannot point immediately to an intercarrier-sound transistor, to an AGC transistor, nor even to the emitter, base, and collector terminals of various transistors.

Instead, the troubleshooter must follow some preliminary map-out procedures before he can say what a particular transistor, diode, or integrated circuit is supposed to do. This is an approach that is quite different from looking at a service manual in which every component, device, functional section, and interconnection detail is set forth, with operating voltages, circuit resistances, and scope waveforms.

It might be supposed that troubleshooting without service data would be comparatively easy with the aid of a wide-band triggered-sweep scope—and this may be true, but not in all circumstances. For example, when the troubleshooter must cope with a "dead" consumer electronic product (without service data), he cannot use a scope, because there are no waveforms available for checking. In this situation, the technician needs various new "tricks of the trade," just to get started on the job.

To get the feel of troubleshooting without service data, the technician must depart from his established perspective, and start with a "black-box" viewpoint; that is, elementary questions, that ordinarily have obvious answers, must now be checked out by specific tests and measurements—some of which are new and unexpected. The following chapters explain how you can build your own specialized testers for troubleshooting without service data.

Hands-on experimental projects are described in the text for reinforced learning, and to show how "without-service-data" troubleshooting expertise can be acquired without difficulty. Reinforced learning can be easy and interesting—even exciting when you see the speed with which you are progressing.

Virtually all electronic troubleshooters agree on the basic principles that time is money and that knowledge is power. Your success in the profession of electronic troubleshooting is limited only by the horizons of your technical know-how. The novel tricks of the trade, new techniques, and troubleshooting approaches described and illustrated in this unusual book provide key stepping stones to take from your present position to your goal.

Robert G. Middleton

Contents

1

Audio Troubleshooting Techniques

*Preliminary Trouble Analysis * Quick Checks * Temperature Quick Checks * Resistor, Capacitor, and Transformer Temperatures * Buzz-Out of a Transistor * Follow-up DC Voltage Measurements * Finger Test * "Blind" Troubleshooting Procedures * In-Circuit Transistor Testing * Malfunction Due to Error in Transistor Replacement * Experiment*

PRELIMINARY TROUBLE ANALYSIS

Preliminary trouble analysis of an audio amplifier can be made as follows:

1. Encourage the amplifier owner or operator to tell you all that he can about his previous experience with the amplifier, and how the present trouble symptoms developed. While listening to his story, turn the power switch on. If there is sound output of any sort, reduce the volume-control setting.
2. As the amplifier operator continues, make a visual inspection of the interconnecting cables to determine if a cable might be unplugged. When the opportunity occurs, ask the operator if he has changed any of the cable connections. Ask him or her to show you anything that may have been changed, either before or after the trouble symptoms developed.
3. Ask the amplifier operator whether another technician (or some friend) may have previously attempted to fix the trouble, and try to obtain all incidental details. While the conversation proceeds, pull and flex the interconnecting cables. (When you least expect it, you may discover a poor

contact to a connector, or a conductor broken inside the cable insulation.)

4. If the system comprises separate preamplifier and power amplifier units, check each unit separately—the fault will almost always be localized to one unit.[1]
5. External visual inspection should be followed up by a close internal visual inspection. Open the cases or cabinets and look for obvious trouble clues. For example, a "new" amplifier was seen to have a diode with a pigtail sprung up at one end due to a cold-solder connection that had let go.[2]
6. With the power switch turned on, advance the volume (loudness) control to maximum, and listen for noise output from the speaker. If there is no discernible noise output, the output section is probably defective. On the other hand, if there is normal noise output, the trouble is most likely to be found in the input section of the amplifier.
7. Measure the power-supply input resistance (at the power-cord terminals, or at the battery connector terminals). This quick check is most informative in case you have a similar amplifier in normal working condition for comparison purposes. Here are some "ball-park" resistance values:

   Realistic stereo preamplifier: 1500 ohms.
   Archer minispeaker-amplifier (battery powered): 7500 ohms with ohmmeter test leads in one polarity; 579 ohms with ohmmeter test leads in the other polarity.
   Realistic SA-10 stereo amplifier: 143 ohms.
   Realistic MPA-20 public-address amplifier: 8 ohms.
   Realistic SA-102 stereo amplifier: 75 ohms.

8. Measure the input resistance of the amplifier. Most amplifiers have series-capacitor input circuitry. In turn, if the input capacitor is leaky or shorted, this fault will show up in a resistance check.

---

[1]In the case of a stereo system, it is often feasible to interchange preamplifiers, power amplifiers, and speakers.

[2]Case History: A stereo preamp was "stone dead." Visual inspection quickly revealed a V_{CC} lead with a broken end.

In virtually all cases, an electrolytic input capacitor is employed. Accordingly, the input circuit is polarized, and two different ohmmeter readings will normally be obtained, if the capacitor is in normal condition. For example, if you use a 50,000 Ω/V meter on its Rx10k range, observe the following responses:

(A) Does the ohmmeter reading "crawl," and eventually indicate infinity? If so, the input capacitor is in good condition, and the ohmmeter is being applied in proper polarity.

(B) If the ohmmeter test leads are now reversed, the ohmmeter reading will "crawl," but the final reading will be short of infinity—for example, 400,000 ohms.

(C) If the ohmmeter reading "crawls" and the final reading is short of infinity, in BOTH ohmmeter polarities, the troubleshooter concludes that the electrolytic input capacitor is leaky, and should be replaced.

9. Note that some amplifiers utilize a shunt input resistor prior to the series input capacitor; in such a case, the foregoing input resistance check cannot be made. (The ohmmeter will immediately indicate the value of the shunt input resistor, and the reading will not crawl.) Note also that when a series input capacitor is checked with an ohmmeter, it acquires a charge—the input terminal should be short-circuited to ground before making an ohmmeter check with reverse polarity.

"BALL-PARK" TEMPERATURES[3]

An informative quick check of an audio amplifier is provided by temperature measurements of devices and components with a temperature probe or with a differential temperature checker. This quick check is most useful if you have a similar amplifier in normal working condition, for comparison purposes. However, even if you do not have a comparison unit available, much useful test data can be obtained.

For example, observe the typical normal temperatures for a driver transistor, and for a small power-type transistor, shown in

[3]Caution: Power-type transistors can run hot enough to burn your finger.

Figure 1-1. With an ambient temperature of 18°C, the body of the driver transistor operates at 27°C; in a stereo amplifier, the driver transistors in the L and R channels will normally operate within 2° or 3° of each other—a substantial difference in temperatures points to a circuit defect.

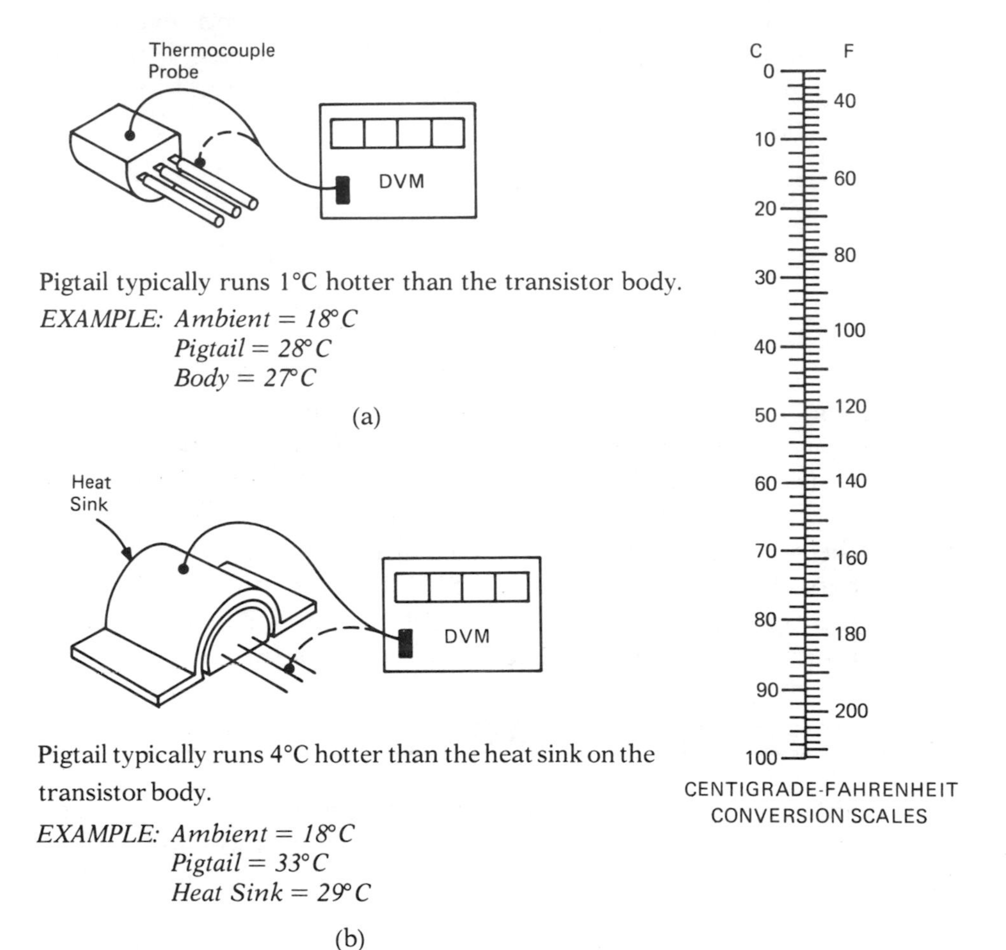

The hottest spot in a transistor is inside at its collector junction. On the exterior of the transistor, a temperature probe will show that the pigtails run hotter than the body of the transistor

Figure 1-1 Example of transistor operating temperatures. (a) driver transistor; (b) small power-type transistor.

Power transistors normally run hotter than low-level transistors; for example, the power transistor exemplified in Figure 1-1 operates at a heat-sink temperature of 29°C. Output transistors in the L and R channels of a stereo amplifier will normally operate within 3° or 4° of each other—a substantial difference in temperatures points to a circuit defect.

Subnormal operating temperatures indicate circuit malfunction, just as abnormal operating temperatures indicate circuit malfunction.

Note that the normal operating temperature of an output transistor (in particular) depends significantly upon whether it is tested *with signal*, or *without signal*. In Figure 1-2, observe that with no signal input, the heat-sink temperature of the output transistor is 28°C, whereas with signal input for a rated output of 2.5 watts, the heat-sink temperature is normally 44°C.

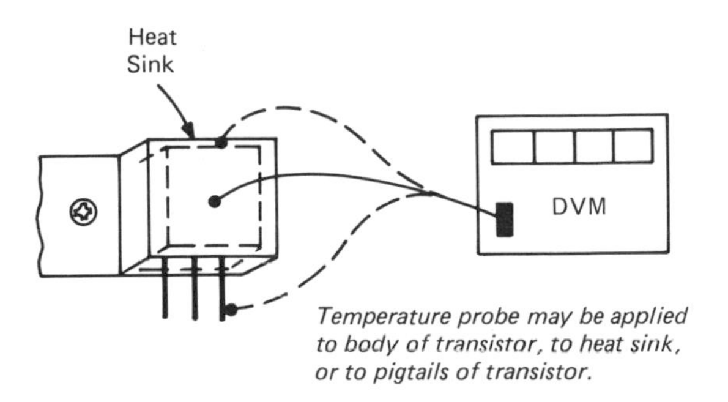

Example: In normal operation, with no signal input, and with an ambient temperature of 20°C, the body temperature of the transistor is 28°, the heat sink temperature is 28°C, and the pigtail temperature is 33°C. Again, in normal operation, with signal input for 2.5 watts output, the body temperature is 44°C, the heat-sink temperature is 44°C, and pigtail temperature is 46°C.

Note: Power output is equal to E^2/R, where E is the r.m.s. voltage of a sine-wave signal across the load, and R is the value of the load resistor. For example, 4.47V r.m.s. across an 8-ohm load is equal to 2.5 watts.

Figure 1-2 Example of output transistor operating temperatures in a 2.5-W audio channel.

Another practical example is shown in Figure 1-3. Here, a power-type integrated circuit is the output device. When checked *without signal,* the heat-sink temperature is normally 25°C. On the other hand, when checked *with signal* for a rated output of 10 watts, the heat sink normally operates at 61°C, or over 140°F.

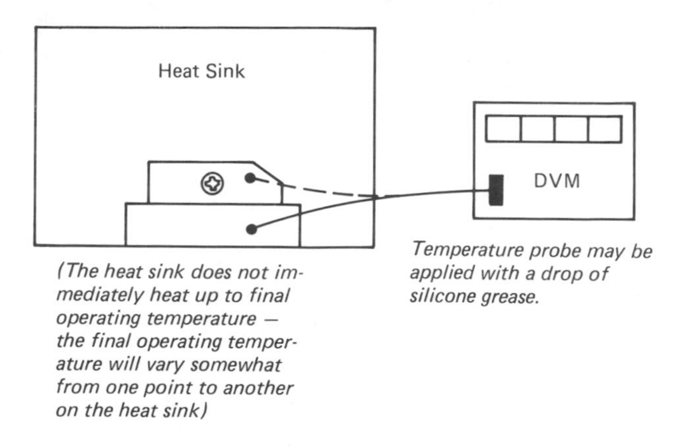

Example: An integrated circuit with a rated power output of 10 watts normally idles (no signal input) at an ambient temperature of 19°C with an IC body temperature of 24°C, and a heat-sink temperature of 25°C. When operated with signal input for an output of 10 watts, the IC body temperature normally increases to 62°C, and the heat-sink temperature increases to 61°C (typical).

Although comparative temperature measurements are most informative when *identical* ICs are checked on a temperature probe, useful data can also be obtained by making comparative temperature measurements on nonidentical ICs that are rated for the *same power output.*

Figure 1-3 Temperature check of power integrated circuit.

RESISTOR, CAPACITOR, AND TRANSFORMER TEMPERATURES

In some trouble situations, highly informative preliminary checks can be made by measuring the temperatures of resistors, capacitors, and transformers. As exemplified in Figure 1-4, each component has its normal operating temperature. This temperature may be the same with signal or without signal, or, it may change from zero signal to full rated output signal level.

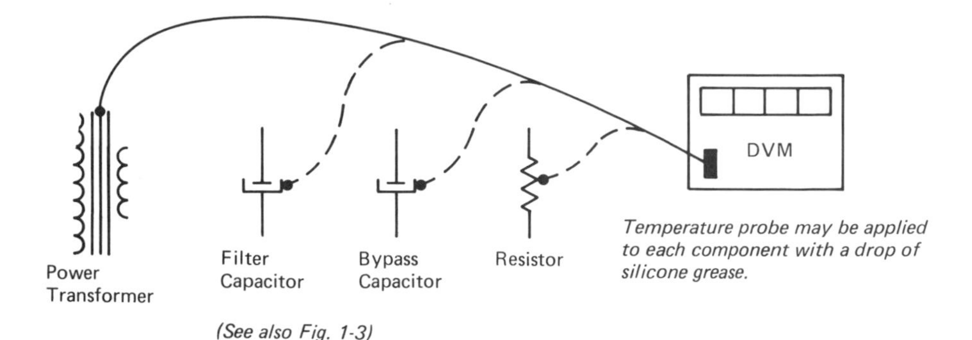

Example: A preliminary checkout of R, L, and C components can be made with a temperature probe while the electronic unit is idling (no signal input). Although certain variations can be expected, these components normally operate at approximately the same temperatures in any unit of electronic equipment. (This is an example for a 20-watt PA amplifier.) With an ambient temperature of 20°C, "ball-park" temperature values for R, L, and C components are typically:

Power transformer, 24°C
Power-supply filter capacitor, 23°C
Electrolytic bypass capacitor, 22°C
Resistors, from 22° to 26°C

Substantially higher temperatures point to an associated malfunction that causes excessive current drain.

Subnormally lower temperatures point to an associated malfunction that causes insufficient or zero current flow.

Figure 1-4 Practical examples of normal operating temperatures for R, L, and C components.

Additional helpful test data can be obtained if you are troubleshooting a stereo amplifier, because corresponding components in the L and R sections *normally* operate at practically the same temperature. Therefore, if a comparative temperature check shows that a given component in the L section is operating at a significantly higher or lower temperature than the corresponding component in the R section, the troubleshooter concludes that a malfunction is occurring.

Whether the malfunction is associated with the L component or with the R component is determined by these related facts:

1. If it has been previously established that the trouble symptom is localized to the L channel (for example), it is reasonable to

conclude that the off-temperature component in the L channel is associated with circuit malfunction.

2. An "out-of-the-ballpark" temperature in itself is a definite indication of circuit malfunction—for example, a small resistor operating in the 60°C temperature range (instead of the 20°C range), points to circuit malfunction in either a stereo or mono amplifier.
3. When both the L and R channels exhibit trouble symptoms, and the temperatures of corresponding components are different (although "in the ballpark"), additional tests are required.

High-Speed Differential Temperature Checker

This quick checker (Figure 1-5a) shows whether a pair of corresponding devices are operating at the same temperature, or at different temperatures. For example, one diode may be placed on a reference IC, and the other diode may be placed on a corresponding suspected IC. If both ICs are operating at the same temperature, the DVM remains zeroed. On the other hand, if the ICs are operating at different temperatures, the DVM will indicate a positive voltage or a negative voltage.

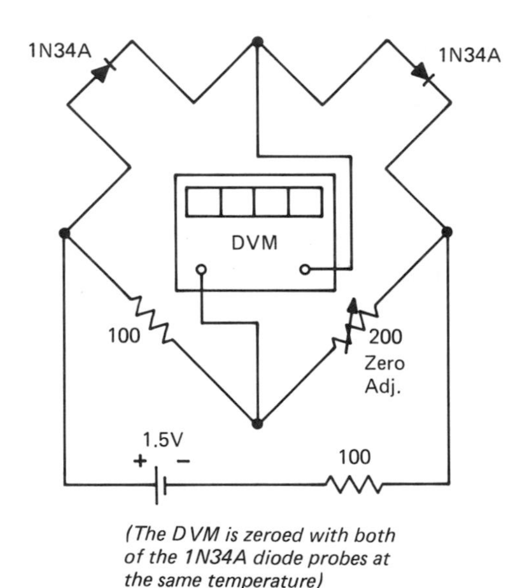

Figure 1-5a

The advantage of the differential temperature probe quick checker is its speed in application. (See Figure 1-5b.) The troubleshooter does not measure individual temperatures, but merely notes whether they are the same, or not. (Whether the DVM indicates zero, or a positive or negative reading.)

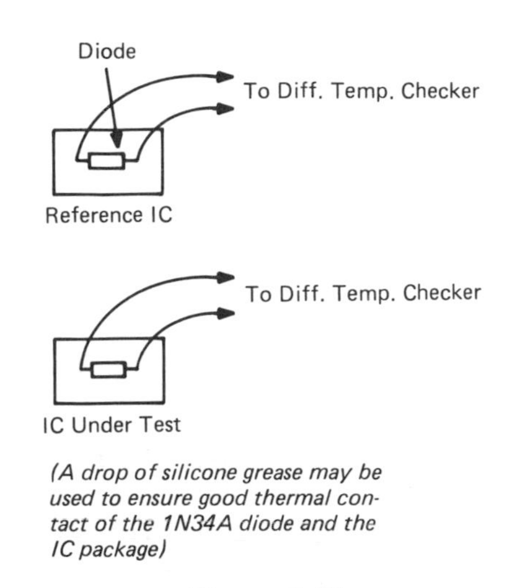

Figure 1-5b

FOLLOW-UP DC VOLTAGE MEASUREMENTS

As shown in Chart 1-1, follow-up DC voltage measurements are generally made to best advantage by buzzing out transistors in stages with malfunctions (or suspected malfunctions). As explained in the chart, the measured voltages may or may not make sense. It is often a question whether nonsense voltages are being caused by a circuit defect or a transistor defect (or possibly both).

The procedure generally is to disconnect the transistor from its circuit for test. (This is ordinarily done by unsoldering the transistor leads from the PC board, but it can also be accomplished by razor-cutting the PC conductors—after the test is completed, the cut conductors can be repaired with a small drop of solder.)

Troubleshooters may use a transistor tester to check out a transistor, or an ohmmeter may be used. Although the following tests are not really new, they are of fundamental importance, and should

CHART 1-1

Buzz-out of a Transistor with a Voltmeter in the Absence of Service Data

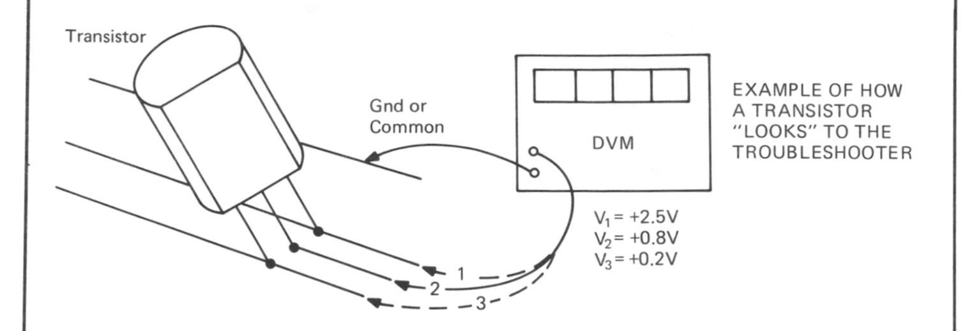

Warning: Avoid accidental short-circuits when measuring voltages!

Buzz-out conclusions: V_2 and V_3 have the smallest difference. Therefore, these voltages identify the base and emitter terminals. In turn, V_1 identifies the collector terminal of the transistor. Since V_1 is positive, this is an NPN transistor. Since most transistors are forward-biased, it is reasonable to assume that V_2 identifies the base terminal, and that V_3 identifies the emitter terminal. This assumption can be verified by making a follow-up shut-off or bias-on test.

The foregoing set of voltage values makes sense and indicates that the stage is workable (in a preliminary analysis). This is just another way of saying that the difference between V_2 and V_3 (0.6V) is a typical forward-bias voltage for a silicon transistor in an amplifier circuit.

Although it is reasonable to assume that the transistor is operating in a common-emitter circuit, and that an emitter resistor will be found between the V_3 lead and ground, this is not always a correct assumption. In other words, this set of terminal voltages could also be found in a common-collector (emitter-follower) circuit, or in a common-base circuit.

It often happens that the measured set of terminal voltages does not make sense. In such a case, the transistor should be removed from its circuit and checked.

The transistor may be checked out-of-circuit with a transistor tester, or with an ohmmeter. Then, proceed as follows:

CHART 1-1 CONTINUED

1. Transistor tests "good."
   Since the transistor is not defective, the troubleshooter now knows that there is a circuit fault present, and that the "nonsense" voltage values are being caused by this fault.
2. Transistor tests "bad."
   Inasmuch as the transistor is defective, it *could* be responsible for the "nonsense" voltage values that were measured. However, there is also a good possibility that there is a circuit fault present that damaged the transistor—and this possibility should be checked before a replacement transistor is connected into the circuit.

   At this time, the troubleshooter should again measure V_1, V_2, and V_3, with the transistor disconnected from its circuit. Then, proceed as follows:
3. Voltages are not "in the ball park."
   Because the terminal voltages still do not make sense, it is logical to conclude that a circuit fault is indeed present. (Transistor bias circuits are usually comparatively stiff, so that the measured bias voltage will be "in the ball park," with the transistor disconnected from its circuit, if there is no circuit fault present.)
4. Voltages are "in the ball park."
   Inasmuch as the terminal voltages make sense in this case, it is logical to conclude that the transistor failed while operating normally, and that the transistor was not damaged by a circuit fault. It is usually safe to replace the defective transistor in this situation. (The chief exception to this rule of thumb is when the circuit has an intermittent fault.)

NOTE: It is often advisable to identify the base, emitter, and collector terminals of a transistor by means of white spots after the terminals have been buzzed out. Thus, a toothpick may be dipped into white poster paint and used to mark the side of the transistor over the terminals. For example, one white spot may identify the emitter terminal, two white spots may identify the base terminal, and three white spots may identify the collector terminal. Thereby, confusion and error are avoided when checking back and forth from one transistor to another.

As a practical note, transistors in the same amplifier do not necessarily employ the same basing. For example, the emitter lead on one transistor might be located in the position of the collector lead on another transistor. It is also quite possible for the base lead on one transistor to be located in the position of the collector lead on another transistor.

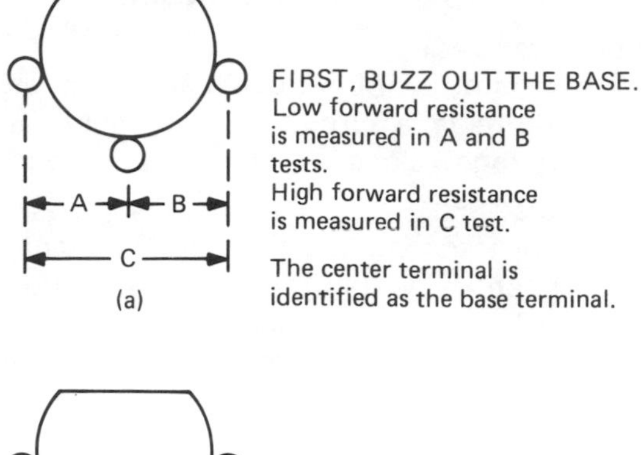

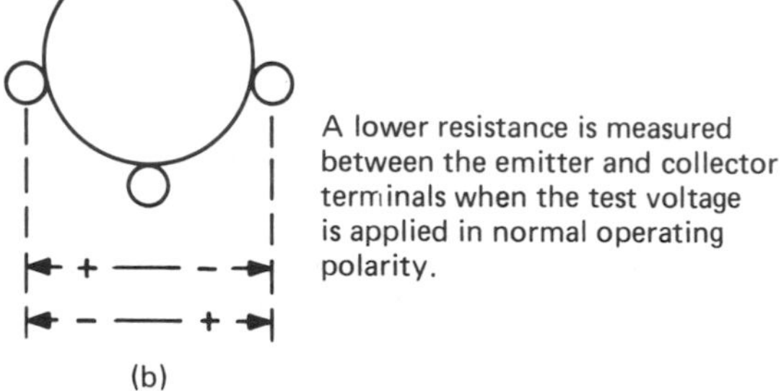

Note: Transistor basing requires attention. When a transistor is replaced, the basing may be different from that of the original transistor, although their appearances are identical.

Case History: An NPN replacement transistor checked out with its collector and emitter terminals reversed, as compared with the original transistor. Moreover, *the basing diagram on the replacement transistor packet was incorrect.* Therefore, the troubleshooter should not assume that the basing of a transistor is the same as would be expected. *Always verify the basing with ohmmeter tests.*

Figure 1-6 Transistor checkout with an ohmmeter. (a) The base terminal has a low forward resistance to each of the two other terminals; (b) a lower resistance is measured between the emitter and collector terminals when the ohmmeter applies a voltage that is polarized as in normal operation. (See also "Finger Test.")

be carefully noted. With reference to Figure 1-6, a transistor is checked out-of-circuit with an ohmmeter as follows:

1. Measure the resistance between each pair of transistor terminals.

2. The two lowest resistance values are from base to emitter and from base to collector, thereby identifying the base terminal.
3. Whether the transistor is a PNP or an NPN type is shown by the polarity of the ohmmeter test leads in measurement of forward resistance.
4. Whether the transistor is a silicon or a germanium type is shown by the value of forward resistance, based on the troubleshooter's experience with the ohmmeter.
5. The collector and emitter terminals can be identified from the rule that a lower resistance is measured between these terminals when the test voltage is applied in normal operating polarity.

Finger Test

Unless the ohmmeter has megohm ranges, a finger test must be used to carry out Step 5. In other words, many ohmmeters cannot indicate the very high resistance between the collector and emitter terminals of a silicon transistor. However, a finger test may be used to provide resistance indication, no matter what kind of ohmmeter is used. To make a finger test, the troubleshooter proceeds as follows:

1. Apply the ohmmeter test leads to the collector and emitter terminals of the transistor (which is collector and which is emitter is unknown at this time).
2. Pinch the base lead and one of the other leads between the thumb and forefinger, to provide "bleeder resistance." Note the resistance reading, if any.
3. Pinch the base lead and the remaining other lead between the thumb and forefinger, to provide bleeder resistance. Note the resistance reading, if any.
4. Reverse the ohmmeter test leads and repeat Steps 2 and 3.
5. The collector is the terminal that provides the lowest resistance reading when its test voltage is bled into the base terminal.

Trick of the Trade: **If your skin is very dry, and you are using a 1000 ohms/volt meter, moisten your fingers slightly to bleed sufficient voltage into the base terminal.**

Caution: When checking circuit boards that have been loosened from the chassis, it is good foresight to cover chassis edges and protruding metal parts with masking tape. This insulation prevents accidental short-circuits to the solder side of the circuit board as it is moved about. The circuit board is usually connected to various front-panel controls and connectors, and to rear-chassis connectors with comparatively short leads. Accidental short-circuits are an occupational hazard.

Case History: A momentary short-circuit between the chassis edge and the solder side of the circuit board blew the regulator transistor in the power supply, which had to be repaired before troubleshooting could continue.

"Blind" Troubleshooting Procedures

After a transistor has been removed from its circuit and tested, it is generally advisable to again measure the circuit voltages, as exemplified in Chart 1-2. In other words, stage malfunction can be caused by transistor failure, by a circuit fault, or by a combination of both. Measurement of circuit voltages with the transistor disconnected can provide essential data concerning the causes of stage malfunction.

CHART 1-2

Basic "Blind" Troubleshooting Procedures

1. When service data is not available, and V_{CC} is present in a malfunctioning audio amplifier, terminal voltages should be measured at transistors in suspect stages. *Example:*

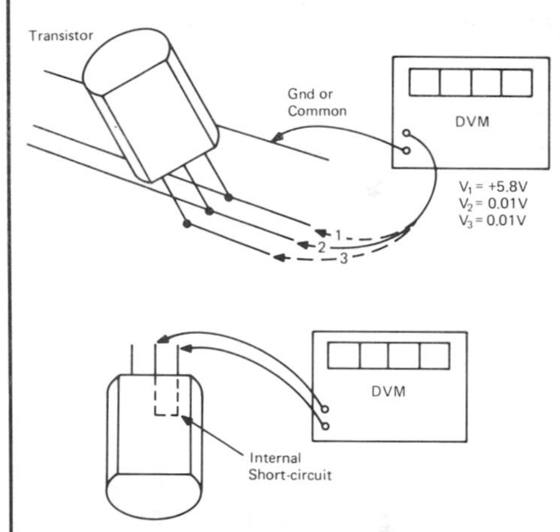

In this example, the DC voltages measured at the transistor terminals in-circuit did not make sense.

Although supply voltage was present, normal bias voltage was not. In other words, V_2 and V_3 would be normal only for a double-ended class-B stage, whereas the transistor under test evidently operates in a single-ended stage. Therefore, the transistor was disconnected from its circuit and checked with an ohmmeter.

CHART 1-2 CONTINUED

As shown in the diagram, an internal short-circuit was indicated by the ohmmeter.

An ohmmeter check of the polarity for the remaining good junction indicated that the transistor was an NPN type. In turn, (1) is the collector terminal, since V_1 is positive.

Before replacing the transistor, the voltages at (1), (2), and (3) were again measured, with the transistor out of its circuit. These voltages were:

$$V_1 = +6.1V \qquad V_2 = +0.84V \qquad V_3 = 0$$

Therefore, it was evident that (2) is the base terminal, and that (3) is the emitter terminal.

The transistor was replaced, and the terminal voltages now measured "sensible" values:

$$V_1 = +4.0V \qquad V_2 = +0.56V \qquad V_3 = +0.01V$$

2. Another example of blind troubleshooting in a malfunctioning audio amplifier is shown below.

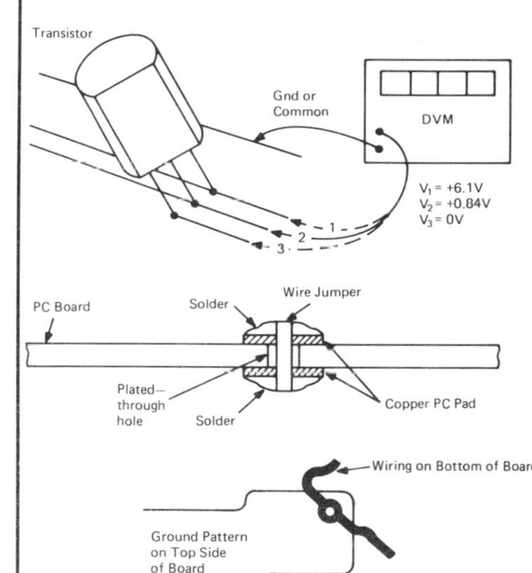

In this example, the DC voltages measured at the transistor terminals in-circuit show that V_{CC} is present, but the bias voltage (0.84 volt) is excessively high.

The transistor was disconnected from its circuit and checked with an ohmmeter; the test indicated that it was a normal NPN transistor.

Next, the voltages at (1), (2), and (3) were again measured, with the transistor out of its circuit. *These voltages were the same as when the transistor was connected into its circuit.*

Therefore, a logical conclusion was that the off-bias voltage was being caused by a circuit defect.

Resistance measurements were made next, with the power turned off. The resistance from (3) to ground measured infinity, or open-circuit.

CHART 1-2 CONTINUED

Visual inspection of the emitter conductor showed that the circuit proceeded via a plated-through hole in the printed-circuit board, and the ohmmeter indicated that the plating was defective. The open circuit was repaired by a wire jumper inserted through the hole, and soldered to the copper pads on the top and bottom of the PC board.

Note: Whenever PC conductors are being checked, or resistance values to ground are being measured, the troubleshooter should use a low-power ohmmeter. This type of ohmmeter will not "turn on" junctions of normal transistors that might be connected into the circuit under test.

3. Another case history of blind troubleshooting in a malfunctioning audio amplifier is shown below:

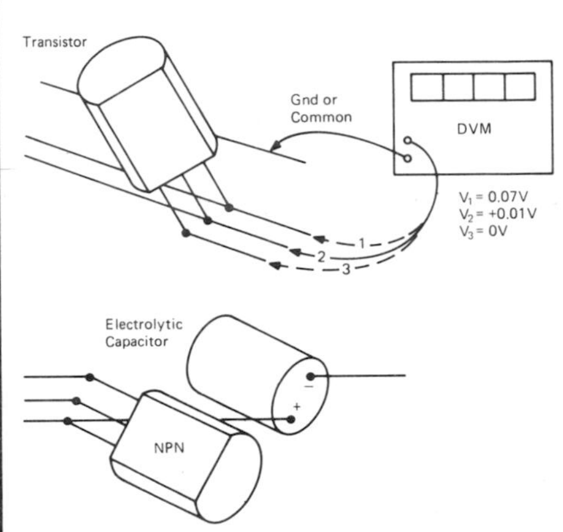

In this case history, the DC voltages measured at the transistor terminals in-circuit show that V_{CC} is greatly reduced and that all terminal voltages are too low for normal operation.

A check of the power-supply voltage indicated that +9 volts was available, although the transistor terminal voltages were greatly subnormal. It was suspected that the transistor had become defective and that it was practially short-circuited.

The transistor was then disconnected from its circuit and checked with an ohmmeter. *Unexpectedly, the transistor was found to be in normal condition.*

Therefore, it was concluded that there was a circuit fault that was "killing" V_{CC}. The voltages at (1), (2), and (3) were again measured, with the transistor out of its circuit, and these voltages were found to be the same as before, when the transistor was connected into the circuit. This fact confirmed the conclusion of a circuit fault.

Since V_1 was the highest of the three voltages, the assumption was made that (1) was the collector terminal of the transistor, and that the circuit fault was likely to be found in this portion

CHART 1-2 CONTINUED

of the PC wiring. Visual inspection showed that an electrolytic capacitor was connected in series with one of the conductors to the (1) terminal.

The DC voltmeter showed that both terminals of the electrolytic capacitor were at a +0.07V potential. This lack of a DC voltage drop across the capacitor indicated that it was probably short-circuited. The capacitor was then disconnected from its circuit and checked with an ohmmeter, and its resistance was found to be nearly zero.

IN-CIRCUIT TRANSISTOR TESTING

An in-circuit transistor tester such as the one illustrated in Figure 1-7 is often useful when troubleshooting without service data, because it identifies all three transistor leads, shows whether the transistor is an NPN or PNP type, and whether it is a silicon or germanium type.[4]

The in-circuit test is valid with circuit shunt resistance as low as 10 ohms, and shunt capacitance up to 15 μF. It is easy to use, because the test clips can be connected in any order to the transistor terminals. The test switch is then moved through its six positions; a pulsating audio tone indicates that the transistor is workable. All three transistor leads are automatically identified.

If the troubleshooter obtains no audio tone indication in a low-drive test, the instrument is switched to high-drive, and the test switch is again moved through its six positions. (High-drive is used when the circuit shunt resistance is quite low.) A workable transistor is indicated by an audio tone, and its base lead is identified.

In the event that the transistor checks "bad" in-circuit, it should be disconnected and rechecked out-of-circuit. A circuit check-out should also be made, as explained above, before the transistor is returned to its circuit, or replaced by another transistor. (See Chart 1-3.)

[4]CAUTION! The amplifier power switch must be turned off, and all capacitors discharged.

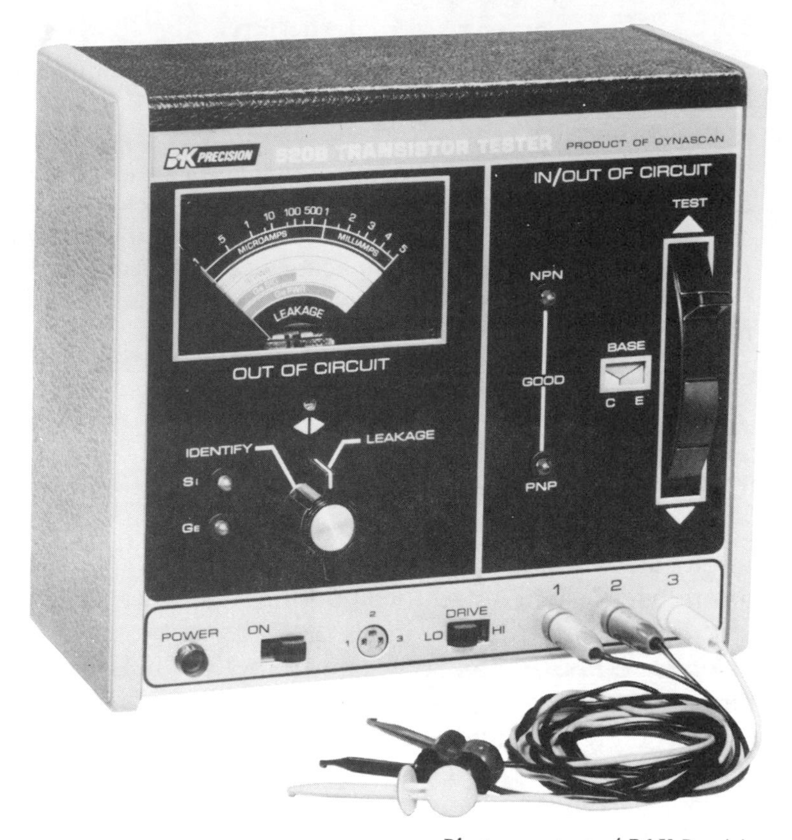

Photo, courtesy of B&K Precision.

Figure 1-7 A widely used in/out of circuit transistor tester

Experiment

The best way to become familiar with modern amplifier circuitry is to construct and check out standard examples, such as the phono preamplifier depicted in Figure 1-8. Although various implementations can be used, it is suggested that the reader "breadboard" the amplifier on an experimenter socket such as the Archer (Radio Shack) No. 276-174.

After the components and devices have been plugged into the experimenter socket, connect the V_{CC} voltage and measure the voltages at the emitter, base, and collector terminals of each transistor with a DVM. If the measured values are significantly

CHART 1-3

Examples of Malfunction Due to Error in Transistor Replacement

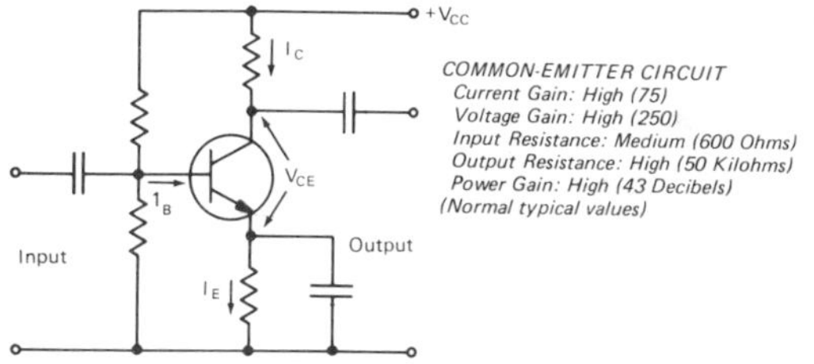
+Vcc
COMMON-EMITTER CIRCUIT
Current Gain: High (75)
Voltage Gain: High (250)
Input Resistance: Medium (600 Ohms)
Output Resistance: High (50 Kilohms)
Power Gain: High (43 Decibels)
(Normal typical values)
Input
Output

Most widely used transistor circuit.

If the collector and emitter leads are accidentally reversed, the stage gain is reduced, typically from 60 to 90 percent, depending upon the type of transistor.

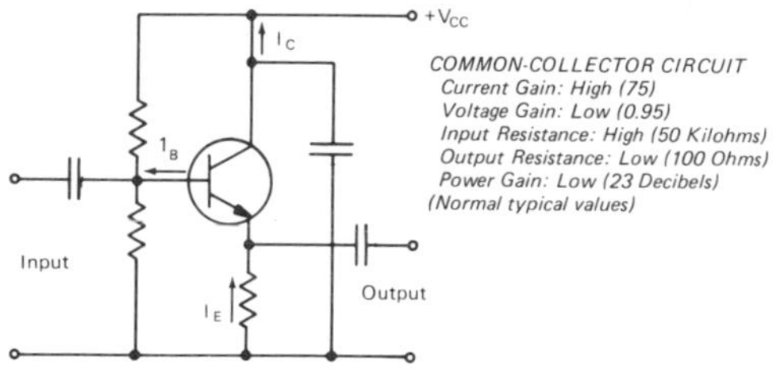
+Vcc
COMMON-COLLECTOR CIRCUIT
Current Gain: High (75)
Voltage Gain: Low (0.95)
Input Resistance: High (50 Kilohms)
Output Resistance: Low (100 Ohms)
Power Gain: Low (23 Decibels)
(Normal typical values)
Input
Output

Second most widely used transistor circuit.

If the collector and emitter leads are accidentally reversed, the stage gain is reduced. In a case history the gain was reduced 40 percent.

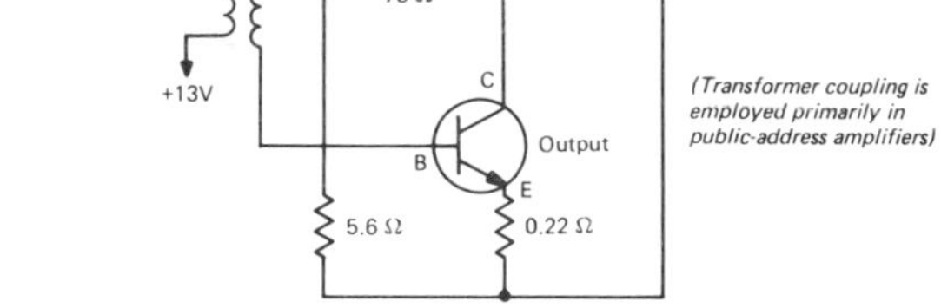
PUSH-PULL OUTPUT CIRCUIT
Base Connections Interchanged
5.6 Ω
0.22 Ω
Output
From Collector AF Pre-amp
75 Ω
−14V
8Ω
Speaker
+13V
(Transformer coupling is employed primarily in public-address amplifiers)

Fundamental output configuration.

If the base leads are accidentally reversed, positive feedback occurs, and a "howl" trouble symptom is likely to occur. When the feedback is insufficient to cause howling, the trouble symptom is serious signal distortion.

different from those indicated in Figure 1-8, check the circuit for error.

Do not disassemble the phono preamplifier at this time; it will be used for a follow-up experiment in the next chapter.

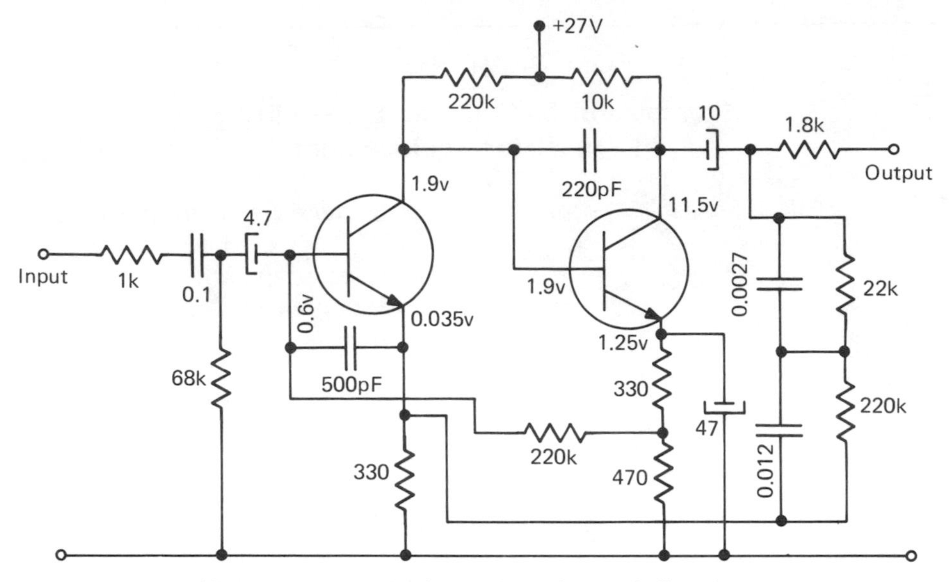

All resistance values in ohms, unless otherwise indicated.
All capacitance values in microfarads, unless otherwise indicated.
Transistors are Radio Shack NPN 276-1603.

PARTS LIST:

(2) 330-ohm resistors, 1/8W
470-ohm resistor, 1/8W
1-kilohm resistor, 1/8W
1.8-kilohm resistor, 1/8W
10-kilohm resistor, 1/8W
22-kilohm resistor, 1/8W
68-kilohm resistor, 1/8W
(3) 220-kilohm resistors, 1/8W
220-pF capacitor
500-pF capacitor
0.0027-microfarad capacitor
0.01-microfarad capacitor
0.012-microfarad capacitor
4.7-microfarad capacitor, 25V
10-microfarad capacitor, 25V
47-microfarad capacitor, 10V
(2) NPN transistors (silicon)
Experimenter socket, Archer (Radio Shack), 276-174
(3) 9-V transistor batteries
(3) battery snaps

Figure 1-8 Experimental phono preamplifier.

FROM "MY OWN NAME" TO "DEAD RECKONING" TROUBLESHOOTING

When troubleshooting, the technician may be completely familiar with the amplifier under test, or the amplifier circuitry may be strange. In other words, the technician may "know the amplifier as well as I know my own name," or he or she may "have never seen anything like this before." In the case of strange circuitry, the technician must proceed by "dead reckoning," or decline to undertake the job.

There are various levels of difficulty to contend with in between these two extremes. For example, the technician may be fairly familiar with the amplifier under test—and he or she may or may not have a similar amplifier in normal operating condition for purposes of comparison.

Again, the technician may have a reasonably similar amplifier in normal operating conditions for purposes of comparison. As an illustration, he or she may have available an earlier or a later production run of the amplifier under test. In turn, he or she may "luck out" and find that he or she has an identical circuit section at hand for comparison tests.

When a technician must troubleshoot strange circuitry by dead reckoning, he or she is thrown completely upon his or her own resources and needs to have in-depth understanding of circuit action. This can be a challenge for the most experienced troubleshooter.

Sometimes a technician happens to know a helpful parts jobber (or even a hobbyist) who is knowledgeable about the particular amplifier under test. In such a case, two heads are better than one, and the troubleshooter should not be too proud to seek advice.

It is also advantageous to keep up with the electronic servicing journals. In various cases, on-target reports of the needed repair procedure will be found.

BASIC AMPLIFIER CLASSIFICATIONS

You will often hear amplifiers described as small-signal or large-signal types, and as low-power or high-power types. From the viewpoint of electrical measurements, these classifications are defined as shown in Figure 1-9. A preamplifier almost always has a

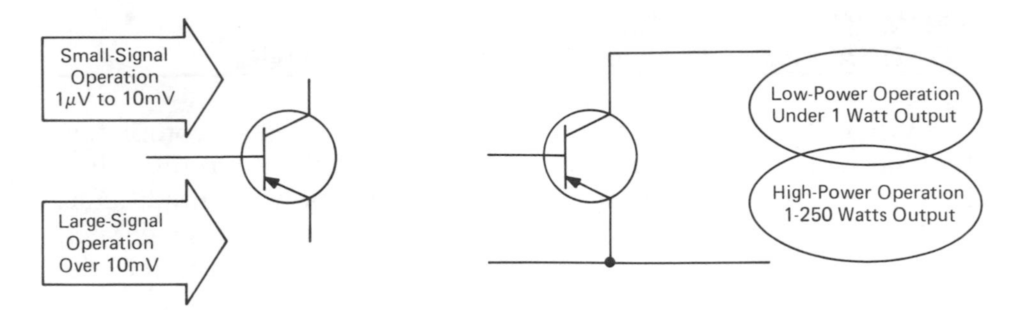

High-fidelity amplifiers differ from public-address amplifiers in that hi-fi units are generally rated for less than 1 percent distortion over a frequency range from 20 Hz to 20 kHz. On the other hand, PA units have a considerably higher percentage of distortion and have a smaller frequency range.

Hi-fi amplifiers are also rated for uniformity of frequency response, such as ±1 dB over the rated frequency range. On the other hand, PA amplifiers have considerably less uniformity of frequency response. PA amplifiers are often categorized as utility amplifiers.

Figure 1-9 Basic amplifier classifications. Power ratings are accompanied by distortion ratings.

small-signal input stage and a large-signal output stage. A power amplifier almost always has a low-power input stage and a high-power output stage.

Playing the Percentages

When you are troubleshooting without service data, the job is greatly facilitated if an identical good unit is available for comparison tests. Even though an identical unit may not be available, a somewhat similar unit may be at hand. Various ball-park comparison tests and measurements can be made on the same sort of unit.

In the event that comparison tests are not feasible, the troubleshooter should first assess the problem, as explained at the beginning of the chapter, and then proceed to "play the percentages." In other words, from a *statistical* viewpoint, it is most probable that a transistor is defective. Electrolytic capacitors are the next most likely suspects, followed by diodes and small fixed capacitors.

Note also that a transistor which operates at comparatively high voltage and/or current is more likely to become defective than is a low-power transistor.

Parts Numbering Sequence

Amplifier circuit boards frequently have parts numbers marked beside devices and components. When troubleshooting without service data, it is often helpful to remember that in virtually all cases, part numbers start with "1" for input components and devices, and then increase sequentially through the network, with the highest part numbers marked beside output components and devices.

TEST EQUIPMENT REQUIREMENTS

It is highly desirable to have a full complement of modern test equipment available. However, not all technicians, experimenters, hobbyists, and students can justify the expense of a large number of lab instruments. The following considerations will serve as a useful guide in this respect.

You will find that a really good DVM is worth its cost, and that it is the keystone requirement. With a top-of-the-line DVM, you can make highly accurate voltage measurements, and you can measure comparatively low voltage values, such as tenths of millivolts. A versatile DVM has both low-power ohms and high-power ohms resistance ranges (plus a continuity beeper), and can measure very high resistance values in nanosiemens units. You will also find that this type of DVM serves as a very sensitive null indicator, and facilitates various test techniques that are impractical with ordinary meters.

A good DVM also provides for accurate temperature measurements and greatly facilitates preliminary troubleshooting procedures. In addition, a top-of-the-line DVM includes basic logic-probe test facilities. This feature can pinch-hit for a digital logic probe, and provides effective intermittent monitoring. Another very practical feature of a versatile DVM is its peak-hold function, which permits the troubleshooter to catch an over-voltage transient, for example.

This is not to say that you might as well go fishing if you do not have a top-of-the-line DVM. One of the things that separates the old timers from the short timers is the ability to make outdated meters perform well. For example, you can measure DC voltages in high-resistance circuits just as well with a 1000 ohms-per-volt meter as with a DVM—*if you use a charge-storage technique with a resistor and large capacitor.* Again, you can easily measure tenths of millivolts with a 20,000 ohms-per-volt meter—*if you use an op-amp preamplifier.* And so on.

After the DVM, your next most important requirement is probably a really good oscilloscope. (Some technicians would say that a lab-type signal generator is the Number 2 requirement—but you will find that a good scope is an accurate frequency indicator for an inexpensive generator.) You will find it worthwhile to invest in a triggered-sweep dual-channel scope with a bandwidth of at least 15 MHz. Practical test work is often facilitated if the scope provides A + B and A− B displays in addition to conventional A & B displays.

The third most important test-instrument requirement (for most electronic technicians) is a good audio oscillator. It should have low distortion (good waveform) to facilitate high fidelity tests. If the audio oscillator provides a choice of single-ended or double-ended (push-pull) output, you can make a comparatively wide variety of tests. It might seem to be a trivial point to note that both line-powered and battery-powered audio oscillators are available. However, there are certain types of test techniques that are facilitated by battery operation (complete isolation of the audio oscillator from the power line).

Again, you can often get by with outdated audio oscillators if certain expedients are utilized. For example, the output frequency can be closely measured with a triggered-sweep scope. The output waveform can be "cleaned up" with a suitable RC filter, and can be further improved with a sharply peaked tone amplifier. A single-ended output can be converted into a usable double-ended output by passing the signal through a high-fidelity audio transformer. This expedient also provides considerable isolation from the power line (avoids capacitive "sneak currents" to the power-line ground).

If your audio oscillator has extended high-frequency output, so much the better. If it also provides a good square-wave output, still better! The basic requirement for good square-wave output is a reasonably fast rise time, such as 20 ns. A square-wave generator not only provides much useful test data in analog circuitry, but also serves as a clock subber and as a signal injector for digital circuit tests.

If you are really involved in digital troubleshooting, you will need a good logic probe, logic pulser, and current tracer. These are the fundamental digital-troubleshooting instruments. They are generally used to localize a trouble area, and follow-up tests with a scope or DVM will sometimes be in order. Unless you are a full-fledged professional troubleshooter, you can dispense with elaborate instruments such as digital multimeters (analyzers), data-domain analyzers, logic clips, and logic comparators.

A good wide-range AM signal generator is a must in any professional shop. Its basic requirements are frequency calibration accuracy, good waveform, and an accurate output level attenuator. If the generator has an FM function, so much the better. However, you will usually choose a separate FM generator with stereo signal output. Also, if you are really interested in high-fidelity troubleshooting, you need a good harmonic-distortion meter.

Other instruments are less essential; for example, you may or may not opt for a versatile semiconductor checker. You may or may not opt for an FM/TV sweep-and-marker generator (although it is a tremendous time-saver). Again, you can get by in color-TV service work without a color-bar generator, but it is a tremendous time saver.

Don't opt for *any* test instrument unless you fully intend to use it on all appropriate jobs. An instrument cannot recoup its cost if it merely gathers dust on the shelf. A thousand-dollar scope is an ill-advised form of window dressing or interior decoration—and your customers can't tell the difference between it and a TV receiver anyhow!

2

Additional Audio Troubleshooting Techniques

*Flexible Audio Signal Tracer * Case History * Self-Oscillatory Quick Check * Sectional Self-Oscillatory Quick Checks * Troubleshooter's Guide List * Series Bias Circuitry * Shunt Bias Circuitry * Thermal Run-Away * Current Feedback * Voltage Feedback * Combined Voltage and Current Feedback * In-circuit Measurement of R and C Values * Experiment * Audio Impedance Checker*

FLEXIBLE AUDIO SIGNAL TRACER

When troubleshooting audio circuitry, a signal tracer such as the one depicted in Figure 2-1 is highly informative. With its aid, the troubleshooter can quickly determine the points in the audio network where signal is present or not present. If signal is present, but is distorted, noisy, or contaminated by hum, these facts are also immediately apparent.

The miniamplifier has high gain, and in turn is very useful as a preamp for a dB meter to check dB gain or loss in low-level circuits. Note that decibel measurements should be made with a 1-kHz sine-wave signal to the audio circuitry under test. On the other hand, when the signal tracer is being used as an audible indicator, it is preferable to employ a speech or music signal. Thereby, distortion is better evaluated. (A speech or music test signal can be provided by a pocket radio, with a test lead from its earphone jack to the input of the audio circuitry under test.)

***Caution:* Avoid overload of the audio circuitry under test. Overload will result in false distortion when the signal tracer is being used as an audible indicator. Overload will also result in erroneous dB measurements. Similarly, it is important to avoid**

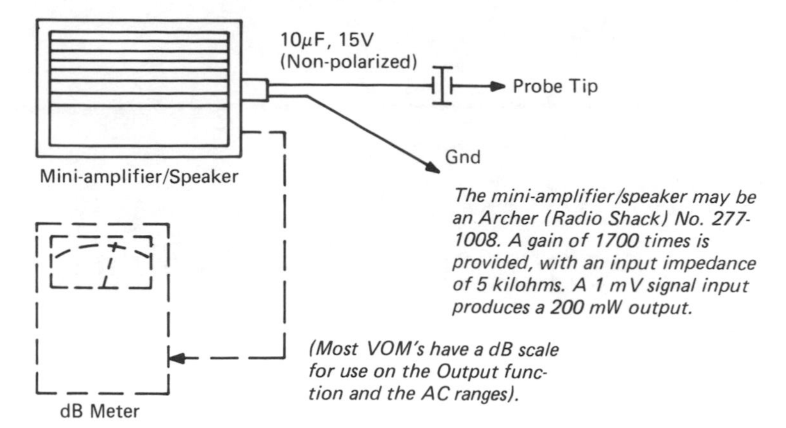

TROUBLE SYMPTOM: SIGNAL IS STOPPED SOMEWHERE IN THE AUDIO SYSTEM

To check dB gain or loss in low-level circuits, a dB meter may be plugged in to the earphone jack on the miniamp housing.

Practical Note: A "sneaky" foul-up that occurs once in a while is caused by intermittent or open-circuited clip leads. In other words, the stranded wire may make poor contact, or no contact, with the test clip. Since the defect is hidden under the insulating jacket, it can cause a tough-dog test problem.

Case History: A packet of 12 new clip leads contained three open-circuited leads.

Figure 2-1 A miniamplifier/speaker operates as a wide-range audio signal tracer.

---

overload of the signal tracer—keep the volume control on the miniamplifier below the point of excessive output and signal clipping.

Observe that a coupling capacitor is included in series with the probe-input lead to the signal tracer in Figure 2-1. This capacitor blocks flow of DC current into the signal tracer from the circuit under test. A nonpolarized coupling capacitor is employed, inasmuch as the troubleshooter will encounter both positive and negative DC voltages in audio test procedures. (See Chart 2-1.)

Case History

As a practical example of audio signal-tracer application, refer to Chart 2-2. In the first case history, the trouble symptom was weak

CHART 2-1

Polarities of Transistor Terminal Voltages with Respect to Ground

"Reasonable" collector voltages: input stage, 2.5 volts; output stage, 15 volts.

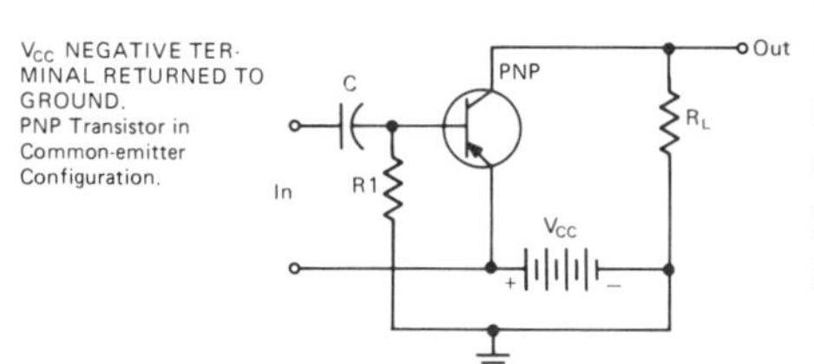

All voltages are positive with respect to ground. (Base voltage is negative with respect to emitter; collector voltage is negative with respect to base and to emitter.)

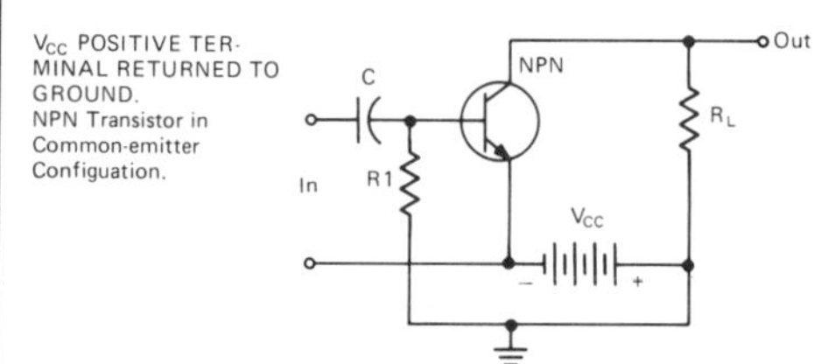

All voltages are negative with respect to ground. (Base voltage is positive with respect to emitter; collector voltage is positive with respect to base and to emitter.)

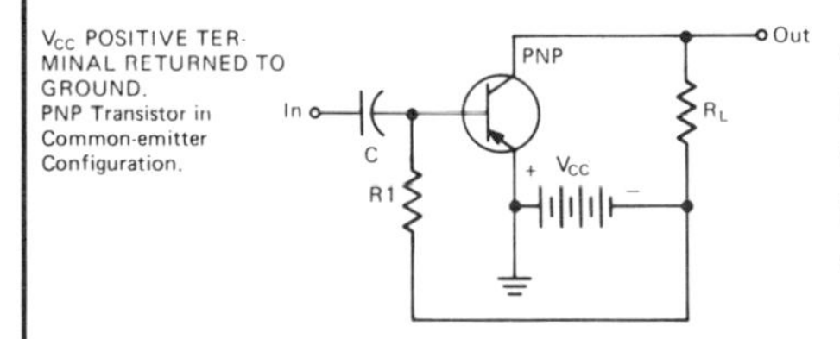

All voltages are negative with respect to ground. (Emitter voltage is positive with respect to base; base voltage is positive with respect to collector.)

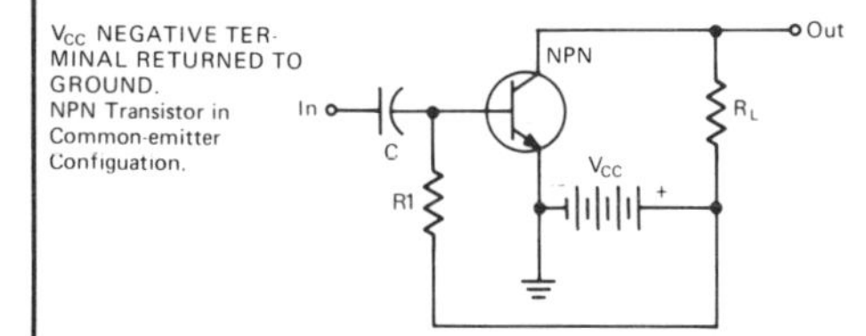

All voltages are positive with respect to ground. (Emitter voltage is negative with respect to base; base voltage is negative with respect to collector.)

sound output. The trouble was tracked down to an open electrolytic capacitor in the negative-feedback loop from Q3 to Q1. *The key trouble clue in this situation is the presence of audio signal voltage across capacitor C1.* Normally, little or no audio signal would be detectable across the bypass capacitor.

CHART 2-2

Case Histories
Single-Ended Audio Output Stages

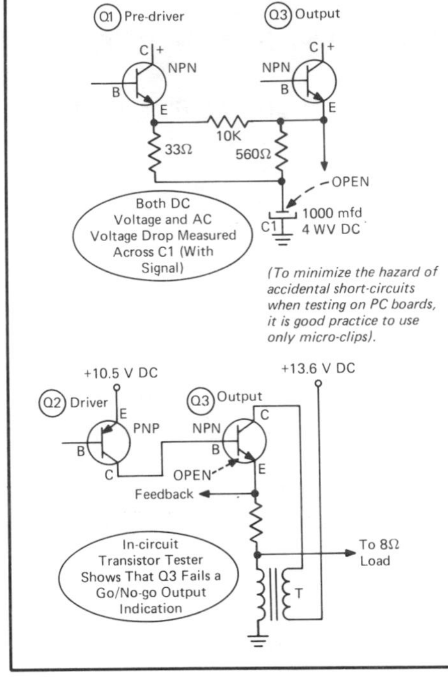

Symptom: Weak sound output. C1 operates in the negative-feedback circuit from Q3 to Q1. (Q2 not shown.) When C1 open-circuits, all of the Q3 emitter signal is fed back to Q1, thereby causing excessive signal attenuation in Q1.

Symptom: No sound output, no audible noise output. When the base-emitter junction of Q3 open-circuits, current flow is stopped in the output circuit. Q3 operates in the emitter-follower mode, and Q2 operates in the common-emitter mode. Transformer T provides auxiliary feedback from collector to emitter of Q3 in a "bootstrap" circuit.

Case histories reproduced by special permission of Reston Publishing Co. and Derek Cameron from Advanced Electronic Troubleshooting.

In the second case history shown in Chart 2-2, the trouble symptom was no sound output, and no audible noise output. *The key trouble clue in this example is the presence of audio signal at Q2, but practically no audio signal in the output circuit of Q3.* These preliminary test data throw suspicion on Q3—the most direct follow-up check is to determine the workability of Q3 with an in-circuit transistor tester.

INTERCHANGE, VOLTAGE-COMPARISON, AND SELF-OSCILLATORY QUICK CHECKS

Interchange quick checks are depicted in Figure 2-2. As shown in the block diagram, when troubleshooting this system without service

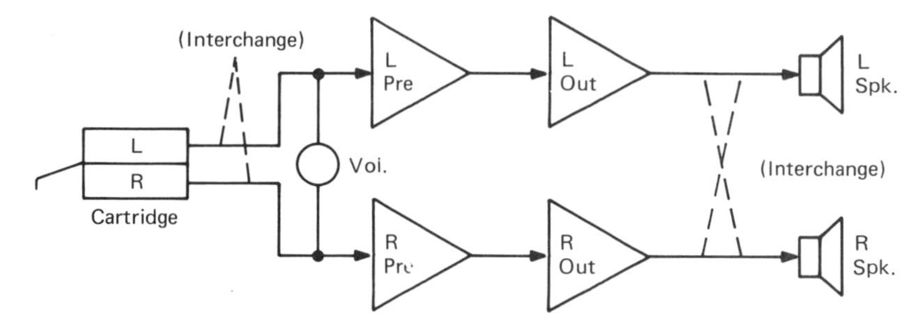

TROUBLE SYMPTOM: ONE CHANNEL "DEAD" OR MALFUNCTIONING

Note: The L and R leads to the cartridge and the L and R speaker plugs can be easily interchanged in most record player systems.

Interchange tests can be followed up with self-oscillatory quick checks of each amplifier and with sectional self-oscillatory quick checks

Illustrative example of comparative L and R preamp DC voltages:

First Preamp Stage
- Collector-to-collector, 0.015V
- Base-to-base, 0.001V
- Emitter-to-emitter, 0.002V

Second Preamp Stage
- Collector-to-collector, 0.020V
- Base-to-base, 0.004V
- Emitter-to-emitter, 0.013V

Comparative DC voltage and resistance checks are very helpful after a defective section is found, in order to pinpoint the faulty device or component. Comparative resistance checks should be made from corresponding test points to ground in the L and R sections, using a low-power ohmmeter.

Figure 2-2 Block diagram of a basic record-changer arrangement.

data, preliminary trouble localization is often facilitated by interchanging the cartridge leads and interchanging the speaker cables.

An interchange test may show, for example, that an observed trouble symptom in the L channel shifts to the R channel when the cartridge leads are switched, or that a trouble symptom that is observed in the R channel shifts to the L channel when the speaker cables are switched. This type of test data provides helpful clues concerning location of the trouble area.

Voltage-comparison quick checks can often narrow down the trouble area and assist in pinpointing a fault. The quick check is made, in the example of Figure 2-2, by connecting a DVM to corresponding transistor terminals in the L and R channels (preamplifier, in this case). When a DVM is connected from collector-to-collector, for example, it measures the difference between the two collector voltages. This difference is normally quite small.

In an ideal situation, the DVM would read zero when connected from collector-to-collector, from base-to-base, or from emitter-to-emitter. However, as set forth in the practical example (Figure 2-2), there will normally be a few millivolts difference in the readings, due to normal component and devices tolerances. On the other hand, a significant voltage reading would point to a fault in the associated circuit.

As also noted in Figure 2-2, resistance-comparison tests made with a low-power ohmmeter from transistor terminals to ground in the L and R sections can often provide further useful test data for pinpointing a fault. (See Chart 2-3.)

A self-oscillatory quick check, exemplified in Figure 2-3, will show whether a section is developing gain, and it is also an informative comparison test to determine whether L and R sections respond in the same manner.

Note in passing that although 220 pF was a suitable value for the positive-feedback capacitor in a typical case history, some other value might be required for quick-checking another amplifier design. In other words, if you do not hear a squealing or motorboating sound from the speaker when a 220-pF capacitor is used, try a 470-pF capacitor, or a 110-pF capacitor.

Sectional Self-Oscillatory Quick Checks

Additional preliminary troubleshooting data can be obtained in many situations by making sectional self-oscillatory quick checks, as exemplified in Figure 2-4. Here, the input stage is a transistor, and the output stage is an integrated circuit.

To check the output stage, a 0.05 μF capacitor is temporarily connected from the "hot" terminal of the speaker to the top of the volume control. In turn, the troubleshooter advances the volume control and listens for a motorboating sound from the speaker. In this example, if the second stage is workable, a motorboating sound is heard from the speaker.

Note that if a motorboating sound is not heard from the speaker when a sectional self-oscillatory test is made in the L channel, a

CHART 2-3

Ohmmeter Indication Errors due to Residual Circuit Voltages

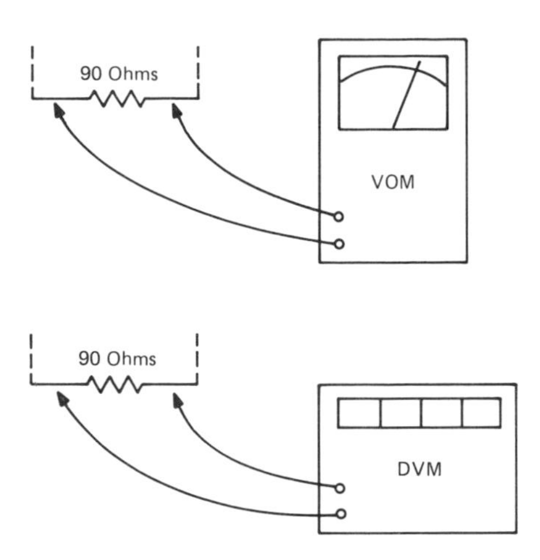

Although an amplifier is turned off, residual circuit voltages may be present due to charges "soaked up" in electrolytic capacitors. Troubleshooters should be on guard against residual circuit voltages, inasmuch as false ohmmeter readings will result.

As an illustration, consider a 20,000 ohms/volt VOM connected across a pair of test points in a circuit, with an actual resistance of 90 ohms existing between the test points. If there is no residual voltage across the test points, the ohmmeter will read 90 ohms.

On the other hand, if there happens to be a residual voltage of 138 mV across the test points, the ohmmeter will give false readings of 60 ohms, or 130 ohms, depending on the polarity of the residual voltage.

As another example of ohmmeter indication error due to residual circuit voltages, consider a 50,000 ohms/volt VOM connected across a pair of test points in a circuit, with an actual resistance of 90 ohms existing between the test points. If there is no residual voltage across the test points, the ohmmeter will read 90 ohms.

On the other hand, if there happens to be a residual voltage of 138 mV across the test points, the ohmmeter will give false readings of 87 ohms, or 115 ohms, depending on the polarity of the residual voltage.

CHART 2-3 CONTINUED

Still another example of ohmmeter indication error due to residual circuit voltages is noted here for a DVM. Consider a DVM connected across a pair of test points in a circuit, with an actual resistance of 90 ohms existing between the test points. If there is no residual voltage across the test points, the ohmmeter will read 90 ohms on either its low-power ohms function, or on its high-power ohms function.

On the other hand, if there happens to be a residual voltage of 138 mV across the test points, the ohmmeter will give false readings of 80 ohms, or 281 ohms on its high-power ohms function, depending on the polarity of the residual voltage. The ohmmeter will give false readings of 10.6 kilohms or 13.6 kilohms on its low-power ohms function, depending on the polarity of the residual voltage.

Therefore, the troubleshooter should be on guard against the possibility of residual circuit voltages.

cross-check can be made in the R channel. Then, if a motorboating sound is heard, the troubleshooter knows that the corresponding section in the L channel is defective.

Next, consider a case wherein the first sectional self-oscillatory quick check in the L channel shows that this section is workable. To quick check the first stage in the L channel, the troubleshooter feeds back the output from the L speaker to the base of the input transistor via the test capacitor, as shown at (2) in Figure 2-4.

As the volume control is advanced, a motorboating sound will be heard from the speaker, if the first stage is workable. In the event that there is no sound output, the troubleshooter should cross-check in the R channel—if sound output is obtained, he knows that the first stage in the L channel is defective.

TROUBLESHOOTER'S GUIDE LIST OF BASIC TRANSISTOR ARRANGEMENTS

Most technicians are familiar with transistor turn-off and turn-on tests.[1] Note in passing that turn-off tests cannot be made

[1]Apprentice technicians may refer to my book, *New Ways to Use Test Meters* (Prentice-Hall, 1983).

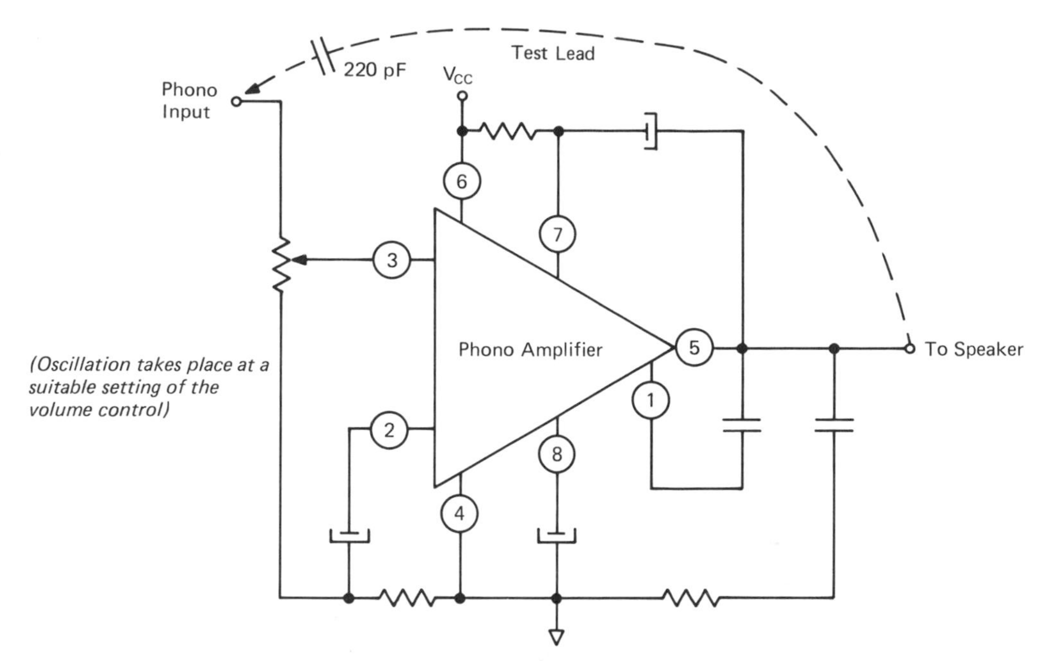

If you are not familiar with the integrated circuits used in consumer electronics products, refer to pages 3 through 110 in *Encyclopedia of Integrated Circuits,* by Walter H. Buchsbaum, Sc.D.

Note: An 8-pin integrated circuit functions as the phono amplifier, in this typical configuration. Positive feedback for a self-oscillatory test is provided by a 220-pF capacitor temporarily connected from the output terminal (5) to the phono input terminal.

Trick of the trade for troubleshooting bypass capacitors

Bypass capacitors can be quick-checked with a DVM; signal voltage must be present, either from an external source, or from a self-oscillation tone signal. The function of a bypass capacitor is to block the flow of DC and to provide an easy escape path for AC. Therefore:

1. More or less DC voltage will normally be measured across a bypass capacitor. If there is no DC drop across the capacitor, it is probably short-circuited.
2. Little or no AC signal voltage will normally be measured across a bypass capacitor. If there is a significant AC voltage drop, the bypass capacitor is probably defective.

Figure 2-3 Self-oscillatory quick check of a phono amplifier.

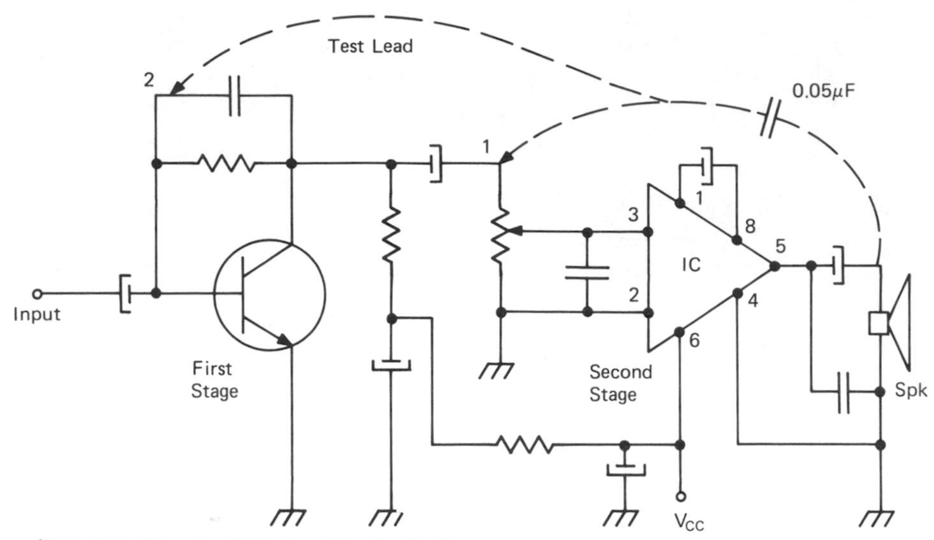

(Note that although a 0.05μF positive-feedback capacitor was suitable in this particular example, a larger or smaller value may be more suitable in other situations).

Note: If the second stage is workable, it will motorboat when the test lead is applied at (1), and the volume control is turned up. If the first stage is also workable, the amplifier will motorboat when the test lead is applied at (2), and the volume control is turned up.

Comparison tests: In the case of a stereo system, self-oscillatory cross-checks of corresponding L and R sections will normally show the same frequency of squealing or motorboating, and the same volume of sound output from the L and the R sections.

Figure 2-4 Example of sectional self-oscillatory quick check.

indiscriminately; appropriate attention must be given to the basic configuration that is being tested. It is particularly essential for the technician to keep the following basic types of transistor circuitry in mind:

1. With reference to Figure 2-5, separate V_{BB} and V_{CC} sources may be encountered on occasion, although this is the exception, and not the rule. Although PNP transistors are shown in the diagram, NPN transistors are also employed, with the polarities of both V_{BB} and V_{CC} reversed. Either turn-off or turn-on tests can be made in these circuits.

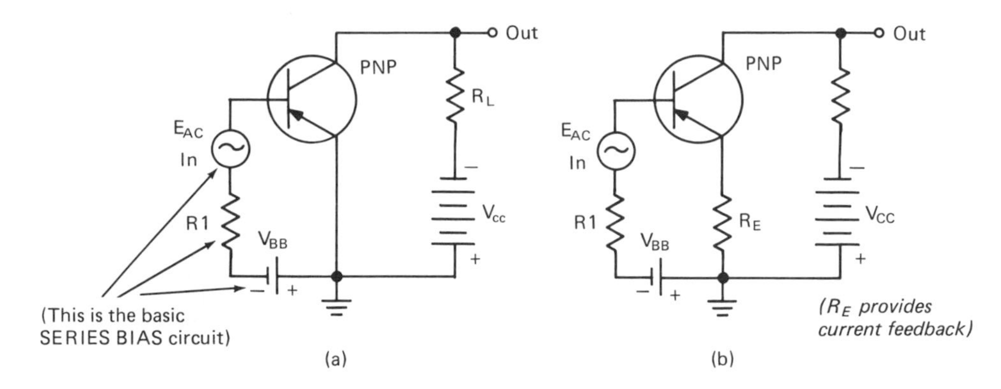

Most CE configurations include an emitter resistor for improvement of bias stability. An emitter resistor also provides current feedback and helps to reduce amplitude distortion.

An emitter resistor, such as R_E, is sometimes called a *swamping resistor*; it masks (minimizes) the change in collector current caused by variations in the emitter-base junction resistance due to temperature changes.

Practical note: R_E provides both DC voltage feedback and AC (signal) voltage feedback. In practice, R_E is often bypassed to "kill" AC feedback while maintaining DC feedback for bias stabilization. (Elimination of AC feedback provides increased stage gain at the expense of increased distortion.) If the bypass capacitor becomes shorted, stage gain remains the same, but bias instability develops. (See also Figure 2-10.)

Figure 2-5 Common-emitter circuits with separate V_{BB} and V_{CC} sources. (a) Simplest arrangement with emitter returned to ground; (b) practical arrangement with emitter resistor.

2. Next, with reference to Figure 2-6, a single V_{BB} and V_{CC} source is exemplified. Observe that, unlike the arrangement in Figure 2-5, a shunt-type bias circuit with a coupling capacitor is utilized with a single V_{BB} and V_{CC} source. Either turn-off or turn-on tests can be made in these circuits.
3. In Figure 2-7, another basic bias arrangement is depicted, wherein a voltage divider is connected in the bias branch circuit. This is also a shunt-type bias circuit with a coupling capacitor—capacitor defects can upset bias-circuit operation. Either turn-off or turn-on tests can be made in these circuits.
4. Observe in Figure 2-8 that the same basic configuration is depicted as was discussed in Figure 2-6—a shunt-type bias circuit is utilized with a coupling capacitor, with a bias resistor

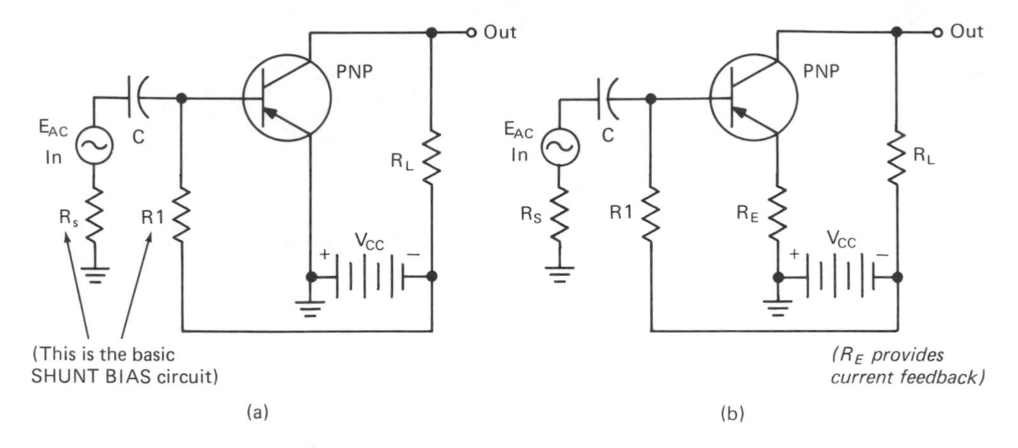

Most CE configurations employ a single voltage source for V_{BB} and V_{CC}. Unless an emitter resistor is included, the bias stability is poor. An emitter resistor also provides reduction of amplitude distortion.

In either of these configurations, a leaky or shorted coupling capacitor can upset the bias circuit and make the transistor look bad. However, the transistor will pass a turn-off or a turn-on test.

Figure 2-6 Common-emitter circuit with a single V_{BB} and V_{CC} source. (a) Simplest arrangement, with emitter returned to ground; (b) practical arrangement with emitter resistor.

connected in the bias branch circuit from collector to base of the transistor. Only a turn-on test is practical—a turn-off test is inconclusive because collector voltage is "bled" away by R_1. Note also that capacitor leakage can upset bias-circuit operation and make the transistor look bad, although the transistor will pass a turn-on test, and will be indicated as workable by an in-circuit transistor tester.

5. Finally, in Figure 2-9, another shunt-type bias circuit is employed with a coupling capacitor, and with a voltage divider connected in the bias branch between collector, base, and ground. We observe that only a turn-on test is feasible in this type of circuit, and capacitor leakage can make the transistor look bad. However, the transistor will be indicated as workable by an in-circuit transistor tester.

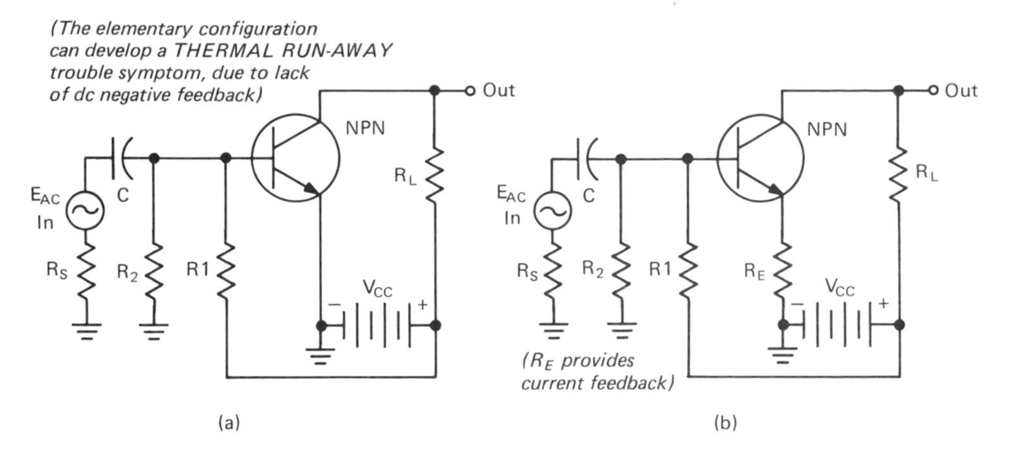

A widely used CE configuration uses a voltage-divider bias circuit. Resistor R_2 has a much smaller value than R_1, and provides reduced resistance from base to ground. Improved bias stability is thus obtained. Additional bias stability is provided by use of an emitter resistor.

In either of these configurations, a shorted or leaky coupling capacitor will disturb the bias circuit and make the transistor look bad. Nevertheless, the transistor will pass a turn-off or a turn-on test.

Note that a thermal runaway trouble symptom consists of excessive and highly distorted output, with the transistor operating at an abnormally high temperature.

Figure 2-7 Common-emitter circuit with voltage-divider bias supply. (a) Simplest arrangement, with emitter returned to ground; (b) preferred arrangement with emitter resistor.

IN-CIRCUIT MEASUREMENT OF R AND C VALUES

Most resistors are color-coded, and their *nominal* values are apparent by visual inspection. Of course, the *actual* value of a resistor may be quite different from its apparent value. Its resistance value can frequently be measured in-circuit with a low-power ohmmeter, provided that it is not shunted by other resistances. If it is shunted by other resistances, one end of the resistor under test must be disconnected for an ohmmeter check.

Capacitors are sometimes marked, and their nominal values are apparent by visual inspection. As in the case of resistors, the actual

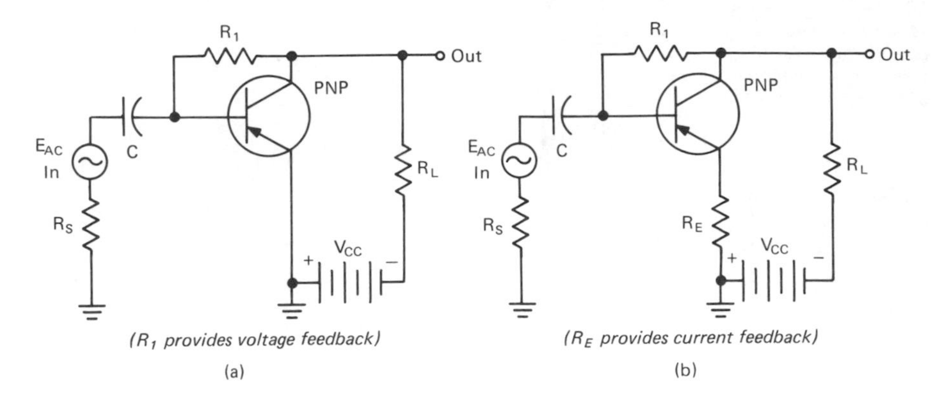

Many CE configurations use negative-feedback bias from collector to base for greater bias stability. An emitter resistor further improves the bias stability. R_1 and R_E also reduce amplitude distortion.

This configuration will not pass a turn-off test, due to bleed of collector voltage through R_1. Only a turn-on test is feasible.

Figure 2-8 Common-emitter circuit with a single V_{BB} and V_{CC} source, and negative-feedback bias. (a) Simplest arrangement with bias resistor from collector to base; (b) preferred arrangement with emitter resistor.

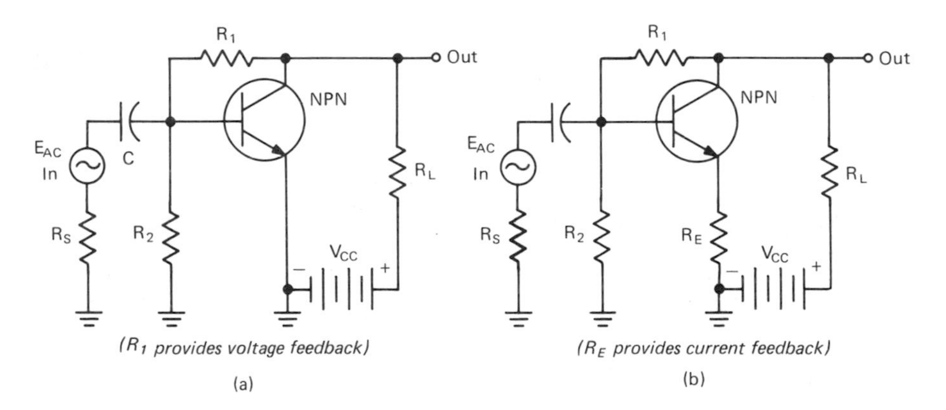

A widely used CE configuration uses a voltage-divider bias circuit, with negative-feedback bias. Both of these features provide increased bias stability. R_1 and R_E also assist in reducing amplitude distortion.

This configuration will not pass a turn-off test, due to bleed of collector voltage through R_1. Only a turn-on test is feasible (or a check with an in-circuit transistor tester).

Figure 2-9 Common-emitter circuit, with voltage-divider bias supply, and negative-feedback bias. (a) Simplest arrangement; (b) preferred arrangement with emitter resistor.

value of a capacitor may be quite different from its apparent value. Its capacitance value can sometimes be measured in-circuit, provided that one end of the capacitor is "floating" (as in an input or output coupling capacitor). Usually, one end of the capacitor is not floating, which requires that one end be disconnected for a capacitance meter check.

Experiment

This experiment employs the phono preamplifier constructed in the preceding chapter, and a low-power ohmmeter. Disconnect the V_{CC} battery. Then, connect a jumper from the $+V_{CC}$ terminal to the $-V_{CC}$ terminal on the preamp. Measure the input resistance, output resistance, and the resistance from each transistor terminal to ground. Compare your measured values with the following typical values:

Input Resistance: Infinite
Output Resistance: 243 kilohms
Input Transistor
  Base: 220 kilohms
  Emitter: 320 ohms
  Collector: 218 kilohms
Output Transistor
  Base: 219 kilohms
  Emitter: 790 ohms
  Collector: 10 kilohms

If your measured values differ significantly from the foregoing typical values, check the circuitry to determine the cause of error.

Do not disassemble the phono preamp at this time; it will be used in another experiment in the following chapter.

AUDIO IMPEDANCE CHECKER

A highly useful audio quick-checker is shown in Figure 2-10. This is an audio impedance checker, used in preliminary troubleshooting procedures.[2] It does not measure AC impedance in ohms—instead, it compares the AC impedance at a test point in a bad amplifier with the

[2]This is a simple tester that responds to total-impedance values. A tester that responds to reactive and resistive components of impedance is depicted in Chart 9-1.

AC impedance at a corresponding test point in a similar good amplifier.

A practical example of application for the audio impedance checker is shown in Figure 2-10. This quick checker is particularly useful for spotting open capacitors, inasmuch as they escape DC voltage and resistance measurements. Note that a capacitor does not have to be completely open in order to be "caught" by the audio impedance checker. For example, if an electrolytic capacitor has lost half of its capacitance, or if it has developed an abnormally high power factor, the associated circuit impedance will be changed, and an audio impedance checker will show this change in a comparative quick test.

Case History: Crossed Leads, False Readings

A hobbyist who was troubleshooting his TV receiver listed the measured DC voltages at numerous test points. None of the values made sense, and some were unbelievably off. The hobbyist called a troubleshooter with whom he was acquainted, and asked him to check out the situation.

At first, it appeared that the hobbyist's VTVM had gone haywire. Then, the troubleshooter noted that the hot lead to the VTVM was black, and that the ground lead was red.

The bottom line was that the hobbyist had accidentally interchanged the connectors while repairing the test leads. Since the 1-megohm isolating resistor was then in series with the return lead to the VTVM, the TV chassis and the meter chassis ground potentials were widely different.

THE "EDUCATED SHOTGUN" APPROACH

When troubleshooting without service data, and when comparison tests cannot be made, the technician must necessarily follow a "shotgun" approach wherein devices and components are replaced at random to determine whether normal operation will resume.

However, this does not mean that the technician must make a totally random attack on the problem. Instead, he can usually troubleshoot with an "educated shotgun." For example, if an integrated-circuit amplifier is dead, the technician will generally try new electrolytic capacitors first. If the amplifier is still dead, he will usually try a new IC next.

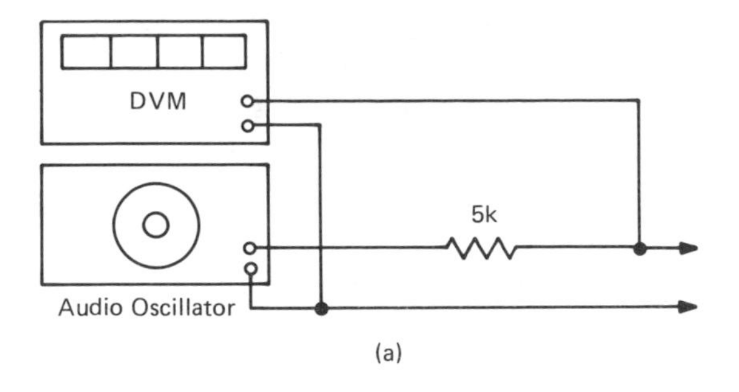

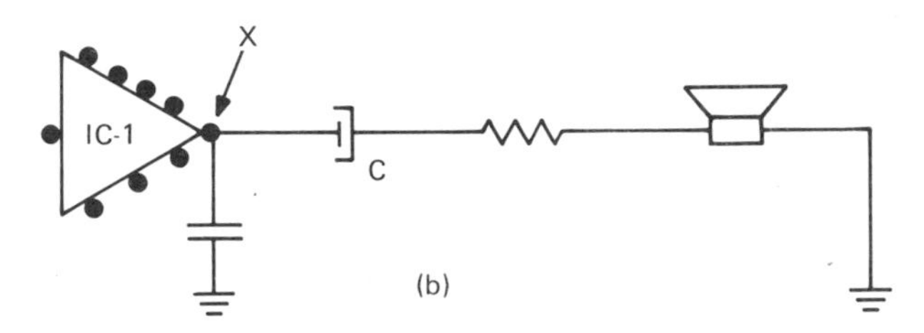

TROUBLE SYMPTOM: NO SOUND OUTPUT

(a) The DVM is operated on its AC voltage function. The audio oscillator is set to 1 kHz with an output level of 2 or 3 volts, as indicated when the output leads are open-circuited.

Practical note: The peak AC voltage from the test point to ground should not exceed 500 mV, in order to avoid junction turn-on and possible confusion of test results.

(b) This is an example of an AF integrated circuit driving a speaker through an electrolytic capacitor and a resistor.

An audio impedance checker is also valuable for testing other types of electronic circuitry.

In this example, there was no sound output from the speaker. All of the DC voltages at the IC terminals made sense on a comparative basis. However, when the audio impedance checker was applied at point X, the DVM indicated that the impedance was much higher in the "bad" amplifier than in the "good" amplifier:

"Bad" Amplifier—56 mV
"Good" Amplifier—5 mV
(DVM indicates 2.75V on open circuit)

The comparatively high AC voltage reading in the "bad" amplifier indicated that the impedance to ground from point X was abnormally high. In this type of circuitry, electrolytic capacitors are ready suspects.

Figure 2-10 Audio impedance checker. (a) Test setup; (b) example of application.

In turn, capacitor C fell under suspicion. When a signal voltage was next applied to the input of IC-1, an AC voltmeter showed signal present on the left-hand end of C, but zero signal on the right-hand end of C.

Therefore, the troubleshooter concluded that C was open-circuited. When a test capacitor was bridged across C, the speaker resumed sound output.

Figure 2-10 (continued)

The next components to be replaced on a trial basis are the fixed capacitors and fixed resistors. As a last resort, the PC conductors may be checked for continuity, on the possibility that a microscopic crack might be present.

Double Trouble Symptoms

From a statistical viewpoint, it is most probable that only one fault will occur in an amplifier at a particular time. However, associated faults ("domino" faults) are not uncommon. For example, a short-circuited capacitor can cause excessive current flow through resistors or transistors, with resulting burnouts.

Note also that comparative tests are not necessarily out of the question simply because the comparison unit is itself defective. As an illustration, an L amplifier may be inoperative because the output transistors overheat and fail. At the same time, the associated R amplifier may be inoperative because of a dead preamp section. In this situation, the L amplifier serves as a valid comparison unit for the R amplifier, and vice versa.

3

Progressive Audio Troubleshooting Procedures

*Color Code Mapping * Signal Flow Diagram * Checkout of Stage Sequence * Residual Bidirectional Characteristic * Self-Oscillatory Quick Check of Single Stage * Check of Thevenin Resistance * Check of Thevenin Impedance * Measurement of Thevenin Impedance in a Low-Level Stage * Basic Types of Audio Power Amplifiers * Distortion Case History * Noise Localization * "Impossible" Case History*

COLOR-CODE MAPPING

When troubleshooting audio amplifiers pinpointing of a fault sometimes requires circuit-action analysis and partial mapping of one or more sections. Because it is easy to forget the location of a particular device or component that has been tested, color-code mapping is often helpful. As exemplified in Figure 3-1, after test data have been noted for a certain transistor, it may be spotted with an identifying color dot.

Similarly, after test data have been noted for a particular electrolytic capacitor, it may be spotted with two identifying color dots. In this manner, the troubleshooter can quickly locate any previously checked component or device.

SIGNAL-FLOW DIAGRAM

Circuit-action analysis often involves signal flow—in other words, is a certain transistor driven by another particular transistor? As the stage sequence is being buzzed out, arrows can be drawn on a signal-flow diagram, as shown in Figure 3-2. In this example, "red"

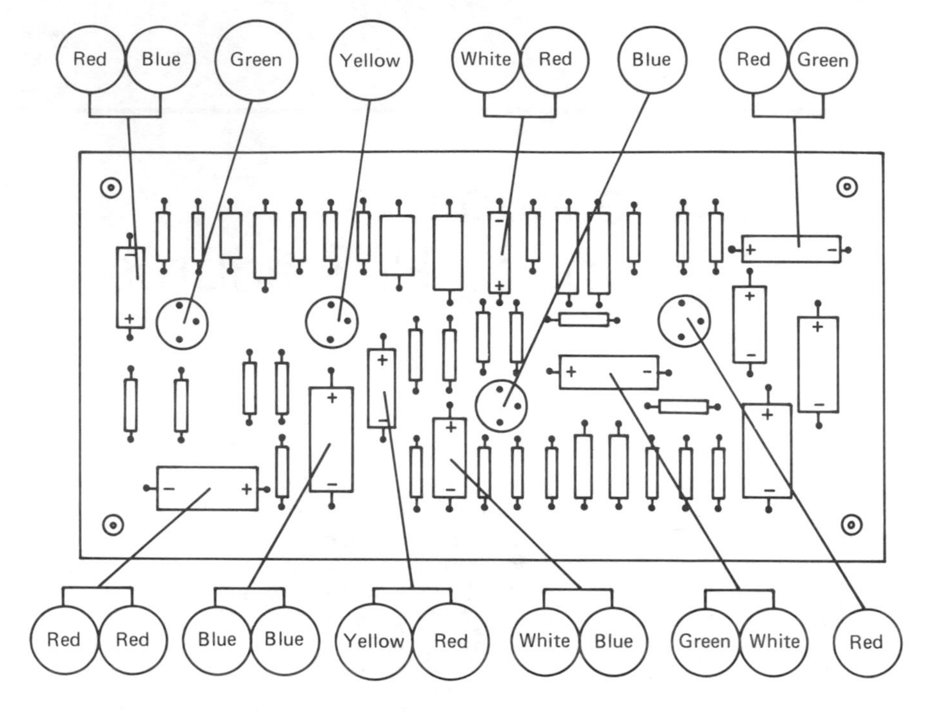

Transistors and electrolytic capacitors are basic "landmarks" in troubleshooting without service data. Accordingly, when voltages are noted, or resistances are noted, the particular component should be spotted with identifying color dots for reference.

Figure 3-1 Color coding of devices and components.

drives "green," "green" drives "yellow," and "yellow" drives both "blue" and "white."

Checkout of Stage Sequence

Stage sequence is easily buzzed out (in normally operating circuits) by means of a noise injector and an audio signal tracer. The procedure is based on the principle that a signal always goes forward through one or more stages—it does not flow backward through a stage—at least in normal operation.

This is just another way of saying that if a noise signal is injected at the output of a stage, a signal tracer applied at the stage input will indicate zero signal, but if the noise signal is injected at the input of the stage, a signal tracer applied at the stage output will indicate a substantial signal level (at least, in normal operation).

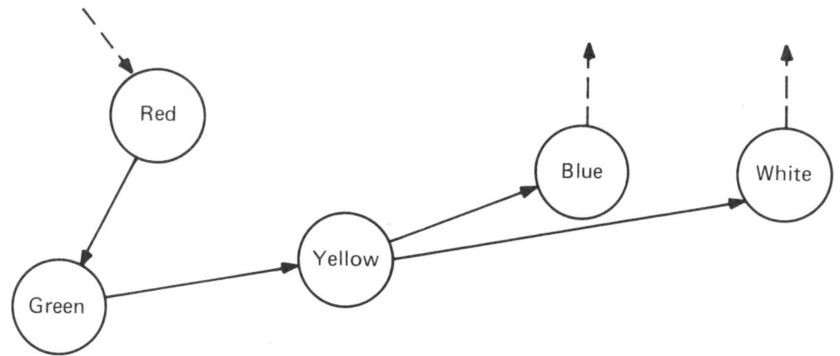

Stage sequence can be checked out with a noise injector and signal tracer. An injected signal normally flows only from a stage input to the stage output—not from an output back to the input.

Color spots can be applied effectively to devices or components with a toothpick dipped in a vial of poster color. For example, the Catalina Tempera poster colors provide bright and permanent color spots.

Quite a bit of printed-circuit tracing will be done visually. Look at the PC board from the component side. Place a bright light behind the soldered side of the PC board. The board is translucent, and the printed conductors are opaque. In turn, it is easy to see through the board and to follow the conducting paths from one component or device terminal to another.

Figure 3-2 Colored transistors may be checked for sequence and marked on the test data sheet.

Residual Bidirectional Characteristic

We usually think of an amplifier as unidirectional in normal operation, and consider that a test signal cannot flow from its output terminals to its input terminals. However, this is not completely true; as exemplified in Figure 3-3, a small fraction of the test signal applied at the output terminals of a stage will be detectable at its input terminals in normal operation.

The troubleshooter should keep this residual bidirectional characteristic in mind to avoid confusion in analysis of circuit action. Note in Figure 3-3 that the collector of the transistor is resistively coupled to the base via the 100-kilohm bias resistor. The collector is also coupled to the base to some extent via junction capacitance. In this example, this residual coupling permits 3 percent of a collector signal to gain entry into the base circuit.

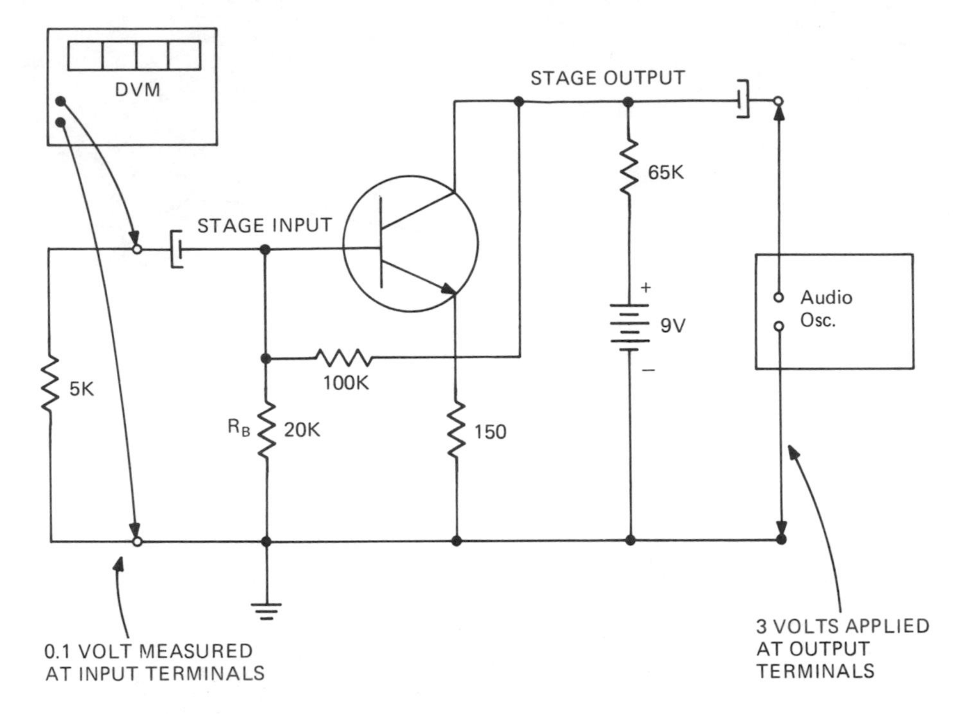

Practical Note: Although the basic transistor amplifier is considered unidirectional, this is not strictly true. As exemplified above, when a 3-volt signal is applied to the normally operating stage, 0.1 volt is measured at the input terminals. This residual bidirectional characteristic should be kept in mind when checking stages for sequence of signal flow.

Figure 3-3 A small fraction of signal voltage applied at the output of a stage normally flows back to the stage input.

SELF-OSCILLATORY QUICK CHECK OF SINGLE STAGE

Previous description was given of a self-oscillatory quick check for an audio amplifier with more than one stage. In this test, a feedback capacitor was temporarily connected from the output to the input of the amplifier system.[1]

This simple test must be slightly elaborated in order to make a self-oscillatory quick check of a single stage. In other words, self-

[1]This check will usually work in a system with more than one stage because cumulative phase shifts will generally provide positive feedback at some audio frequency.

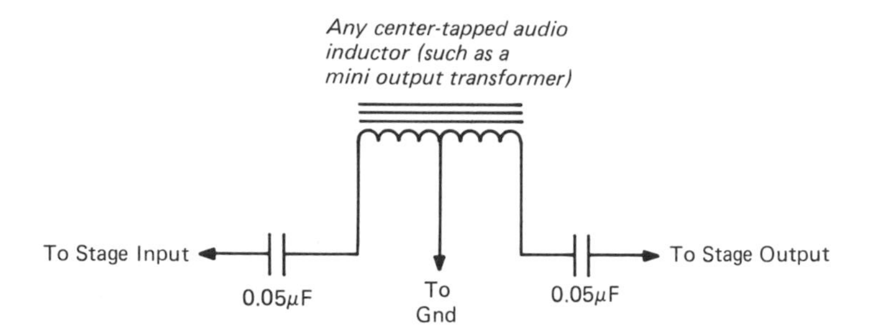

This test gimmick temporarily converts a common-emitter stage into an audio-frequency Hartley oscillator. When used with a signal tracer, it provides a quick check of circuit action (ability of the stage to amplify).

Note that this is an in-circuit gimmick. In other words, it will produce oscillation even though the input and output of the common-emitter stage may be shunted by comparatively low impedances.

Figure 3-4 Positive-feedback gimmick for quick check of a common-emitter stage.

oscillation requires 180° phase shift of the feedback signal in the case of a single common-emitter stage. To obtain this 180° phase shift, a gimmick like the one shown in Figure 3-4 is used. This arrangement provides 180° phase shift by autotransformer action.

In application, a audio signal tracer is connected at either the output or the input of the stage to be checked. The positive-feedback gimmick is connected with one end at the stage output, with the other end at the stage input, and with the center tap returned to ground. If the common-emitter stage is workable, an audio tone will be heard from the signal tracer. Lack of an audio tone indicates that the gain of the stage is very low, or zero.

CHECK OF THEVENIN RESISTANCE

Although the idea of Thevenin resistance measurement is not really new, its application (with incidental test data) is of fundamental importance in troubleshooting without service data. Therefore, suitable Thevenin test procedures are explained below. This method reduces electronic circuitry to its most fundamental equivalent circuit, as seen from the selected test point. As a practical example, if different values of Thevenin resistance are measured for corresponding configurations in L and R channels, the troubleshooter knows that a fault is present in the section under test. (See Chart 3-1.)

CHART 3-1

Check of Thevenin Resistance

Thevenin resistance measurements provide an informative evaluation of the overall condition of an electronic circuit or network, as shown in the diagram.

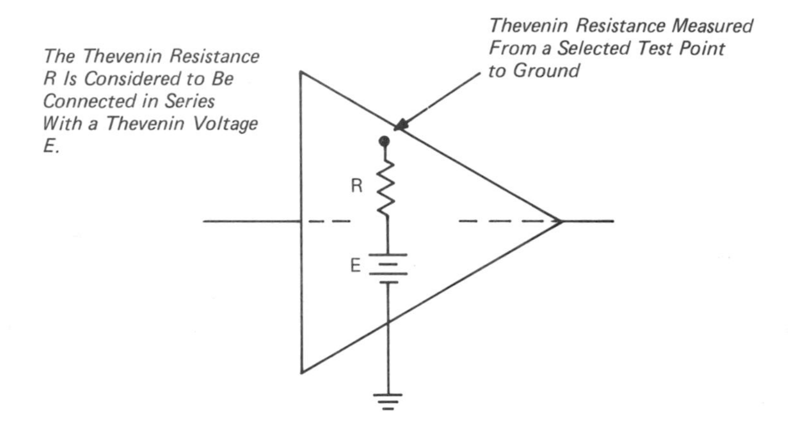

Thevenin's Theorem:

$$I_1 = E/(Z + Z_1)$$

where I_1 is the current in Z_1; E is the open-circuit voltage between any two terminals in a linear network before Z_1 is connected; Z is the impedance between the terminals before Z_1 is connected; and Z_1 is the additional impedance subsequently connected between the terminals. [$(Z + Z_1)$ is replaced by $(R + R_1)$ in a DC circuit.]

A Thevenin resistance measurement is made as follows:

1. The DC voltage is measured from the selected test point to ground.
2. A resistor of suitable value is then shunted from the selected test point to ground, and the resulting decrease in voltage is noted. (This decrease in voltage should be in the range from 10 to 15 percent, and not more than 20 percent.)
3. Ohm's law is then applied to calculate the value of the Thevenin resistance.
4. If the Thevenin resistance measured from a selected test point

CHART 3-1 CONTINUED

in the R channel is the same as that measured from a corresponding test point in the L channel, it is concluded that there is no defect in the associated circuitry.

5. On the other hand, if the two measured values of Thevenin resistance are in substantial disagreement, it is concluded that a defect is present in one of the channels.

As an example of Thevenin resistance measurement, the voltage from a transistor collector terminal to ground may be measured in an amplifier system. If the collector voltage measures 5 volts, and then decreases to 4.5 volts when a 45-kilohm resistor is shunted from the collector terminal to ground, the Thevenin resistance is equal to 5,000 ohms. In other words, the Thevenin resistance is equal to 0.5 volt divided by the current drawn through the shunt resistor, or 0.5/0.0001 = 5,000 ohms. Note that the current drawn by the shunt resistor can be directly measured, if desired, or the current value can be calculated: 4.5/45,000 = 0.0001. Note in the foregoing example that the collector voltage decreased by 10 percent when the shunt resistor was connected from the collector terminal to ground. This 10 percent decrease is within normal circuit-operation tolerances, and does not disturb circuit action appreciably.

With reference to Figure 3-5, a test point is designated at the collector of the input transistor in the amplifier. It is evident that the Thevenin resistance value takes into account a series-parallel network comprising the collector load resistor, the base input resistance of the second transistor, the bias resistor, the input resistance of the first transistor, the output resistance of the first transistor, and the resistance of the emitter circuit for the first transistor.

In turn, the Thevenin resistance value is a very informative quick check. It will indicate (on an L-R comparative basis) whether a transistor may be defective, whether a resistor may be off-value, or whether a capacitor may have substantial leakage. (If the Thevenin resistance is off-value, further tests must be made to pinpoint the faulty device or component).

Consider the measurement of Thevenin resistance in Figure 3-6, from test point T to ground. Since T is above-ground potential, a voltmeter must be used to indirectly measure the Thevenin resistance. The procedure is as follows:

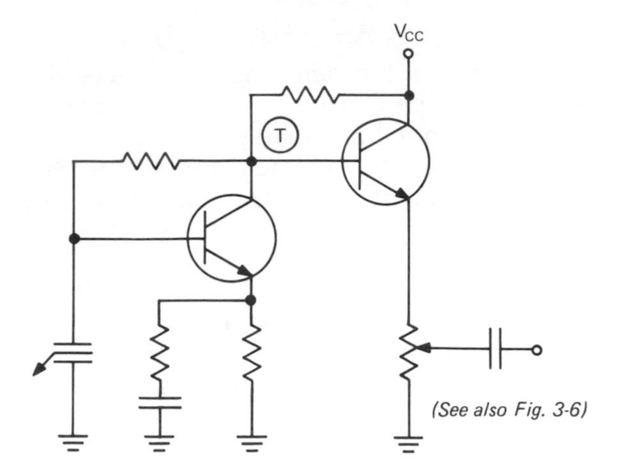

The Thevenin resistance at a test point is very informative because this is a composite resistance value represented by the stage network of fixed resistance and junction resistance values under operating conditions. Serious leakage in a capacitor also shows up.

Figure 3-5 Phono input amplifier; the Thevenin resistance from test point T to ground is measured with a DC voltmeter and a resistor.

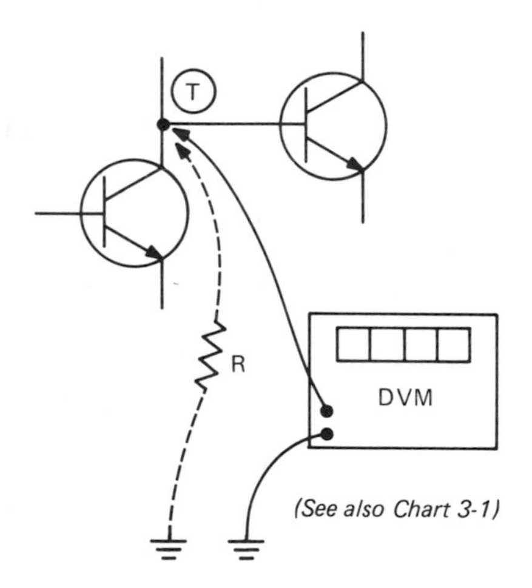

If a DC current meter is available, it may be connected in series with R. Otherwise, the current value may be calculated in terms of the resistance of R and its voltage drop.

Figure 3-6 Test setup for measurement of Thevenin resistance value from T to ground.

1. Connect the DVM between test point T and ground.
2. Measure the DC voltage that is present, with the circuit operating normally.
3. Shunt a suitable value of resistance from T to ground (a value that reduces the voltage from 10 percent to 20 percent).
4. Measure the resulting value of DC voltage.
5. Apply Ohm's Law to calculate the value of the Thevenin resistance.

Example: **The DC voltage at T measures 5.5 volts with the circuit operating normally. When a 0.5 megohm resistor is connected from T to ground, the DC voltage is reduced to 5.0 volts. This is a reduction of 0.5 volt. The current in the test resistor is equal to 5/500,000, or 10 microamperes. Therefore, the value of the Thevenin resistance is equal to $0.5/10 \times 10^{-6}$, or 50 kilohms.**

Thevenin resistance may be measured either *with signal* or *without signal* present. The two values may or may not be the same—either in normal operation or with a circuit defect. In any case, a double check of circuit action is provided by measuring Thevenin resistance with signal, and without signal present.

CHECK OF THEVENIN IMPEDANCE

As in the case of Thevenin resistance, the idea of Thevenin impedance measurement is not really new. However, in view of its basic importance in troubleshooting without service data, this test method is explained below. It reduces electronic circuitry to its most fundamental equivalent AC circuit, as seen from the selected test point. As a practical example, if different impedance values are measured for corresponding L and R configurations, the troubleshooter knows that a fault is present in the associated section. (See Chart 3-2.)

A Thevenin impedance measurement is made with AC signal present. This AC signal may have its source at the input of the amplifier, or, it may have its source at the input of the section under test—an injected AC signal.

Observe that there are basic distinctions between a Thevenin impedance value and a Thevenin resistance value. In other words, when the Thevenin resistance is measured, the voltage source is the

CHART 3-2

Check of Thevenin Impedance

When a check of Thevenin resistance is inconclusive, additional circuit-action test data can often be obtained by means of a quick check of Thevenin impedance, as shown in the diagram.

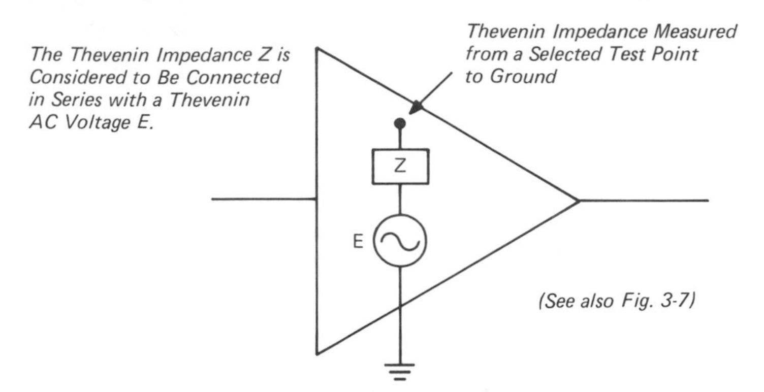

A Thevenin impedance measurement is made as follows:

1. The AC signal voltage is measured from the selected test point to ground.
2. A resistor of suitable value is then shunted from the selected test point to ground, and the resulting decrease in AC voltage is noted. (This decrease in voltage should be in the range from 10 to 15 percent, and not more than 20 percent.)
3. Ohm's law is then applied to calculate the value of the Thevenin impedance.
4. If the Thevenin impedance measured from a selected test point in the R channel is the same as that measured from a corresponding test point in the L channel, it is concluded that there is no defect in the associated circuitry.
5. On the other hand, if the two measured values of Thevenin impedance are in substantial disagreement, it is concluded that a defect is present in one of the channels.

As an example of Thevenin impedance measurement, the AC voltage from a transistor collector terminal to ground may be measured in an amplifier system. If the collector AC voltage measures 4 volts, and then decreases to 3.3 volts when a 291-kilohm resistor (with a 1μF blocking capacitor) is shunted from the collector terminal to ground, the Thevenin resistance is equal to 61.9 kilohms.

In other words, the Thevenin impedance is equal to 0.7 volt

CHART 3-2 CONTINUED

divided by the AC current drawn through the shunt resistor. Or, $0.7/11.3 \times 10^{-6}$ equals 61.9 kilohms.

Note that if your DVM has an AC current function, this current flow can be directly measured, or the current value can be calculated from Ohm's law: $3.3/291{,}000 = 11.3 \times 10^{-6}$.

Observe in the foregoing example that the collector AC voltage was decreased by approximately 18 percent when the shunt resistor was connected from the collector terminal to ground (via the blocking capacitor). This is as large a percentage decrease in signal voltage as should be utilized in practice.

V_{CC} supply. On the other hand, when the Thevenin impedance is measured, the voltage source is at the input of the network under test.

Another basic distinction between a Thevenin impedance value and a Thevenin resistance value is that the resistance measurement takes into account only the resistive paths in the network. On the other hand, a Thevenin impedance value takes into account both the resistive paths in the network and also the reactive (capacitive or inductive) paths in the network.

A Thevenin impedance value is normally considerably greater than a Thevenin resistance value, when measured at the collector terminal of a transistor in a typical amplifier configuration. As an example, a basic amplifier stage exhibited a Thevenin impedance of 61.9 kilohms, whereas the Thevenin resistance at the same test point was 9500 ohms. This wide difference between resistance and impedance values results from the distinctive voltage sources, as noted above.

Thevenin impedance checks are particularly helpful in troubleshooting circuitry that contains capacitors. In other words, an open coupling, bypass, decoupling, or frequency-compensating capacitor will escape a Thevenin resistance check. On the other hand, an open capacitor will often be caught in a Thevenin impedance check (see Figure 3-7).

Example: **The Thevenin impedance at the collector of the transistor in an audio stage measured 40 kilohms in normal operation. On the other hand, when the coupling capacitor to the following stage was open, the measured Thevenin impedance increased to approximately 62 kilohms.**

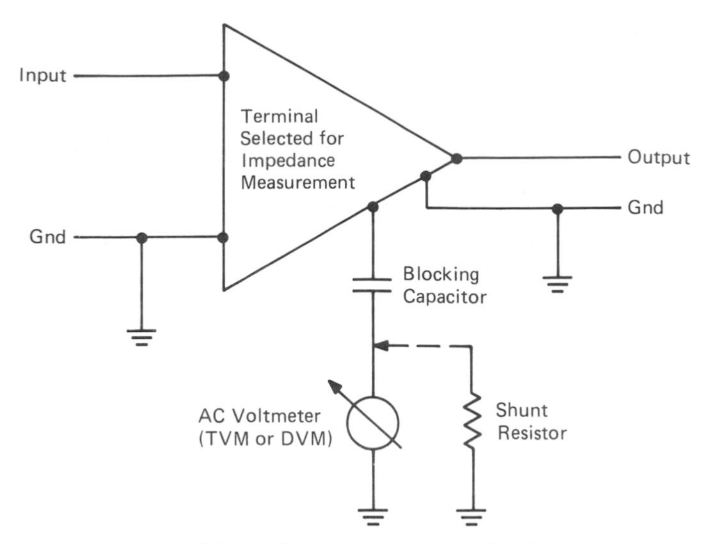

This measurement of impedance gives its total value in ohms, without regard to its resistive and reactive components. A total value serves to reveal serious reactive faults.

Figure 3-7 Test setup for measuring the Thevenin impedance from a test point to ground.

MEASUREMENT OF THEVENIN IMPEDANCE IN A LOW-LEVEL STAGE

A service DVM cannot measure low-level AC voltages, like those encountered in the early stages of a tape-recorder amplifier, for example. In turn, the test signal must be amplified before it is measured. A signal-tracer amplifier may be utilized, as previously noted, or the output section (if operative) will serve the purpose, as depicted in Figure 3-8. The procedure is:

1. Use a prerecorded 1-kHz test tape, or induce a 1-kHz test signal into the play head, as shown in Figure 3-8.
2. Connect an AC voltmeter at the output of the amplifier system.
3. Adjust the input signal level for rated amplifier output voltage.
4. Apply a shunt resistor with a blocking capacitor as shown in the diagram. The shunt resistor should have a value that reduces the output voltage by 10 or 15 percent, but not more than 20 percent.
5. Ohm's law is then applied to calculate the value of the

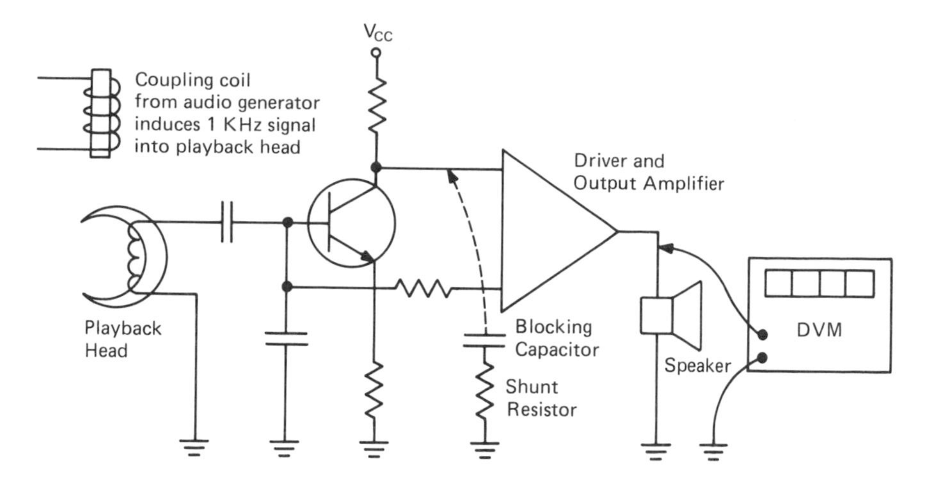

Note: It is permissible to use the driver and output amplifier as an instrument amplifier because the gain factor that is involved cancels out in the equation for calculation of Thevenin impedance.

Figure 3-8 Measurement of the Thevenin resistance at a test point in a low-level stage.

Thevenin impedance at the test point. For example, if the AC output voltage initially measures 2.0 volts, and then decreases to 1.65 volts when a 291-kilohm resistor is shunted to ground, the Thevenin impedance is equal to 61.9 kilohms ($0.35/5.65 \times 10^{-6} = 61.9$ kilohms).

BASIC TYPES OF AUDIO POWER AMPLIFIERS

When you arc buzzing out audio circuits the pieces fall into place faster and more easily if the troubleshooter recognizes basic amplifier configurations, such as the power-amplifier arrangements exemplified in Figure 3-9 and 3-10. The class-B and class-AB complementary-symmetry circuits are in wide use; the bridge configuration is less common.

The common-emitter version of the direct-coupled complementary-symmetry arrangement is also frequently encountered, as is the common-collector version. An increasing number of the more recently designed output amplifiers employ the compound-connected complementary-symmetry configuration. These are all output transformerless (OTL) arrangements. Note, however, that the obsolescent transformer-output configuraton will still be encountered in the majority of public-address amplifiers.

Note: Preliminary troubleshooting of any type of audio power amplifier is often facilitated by making comparative temperature checks on output transistors.

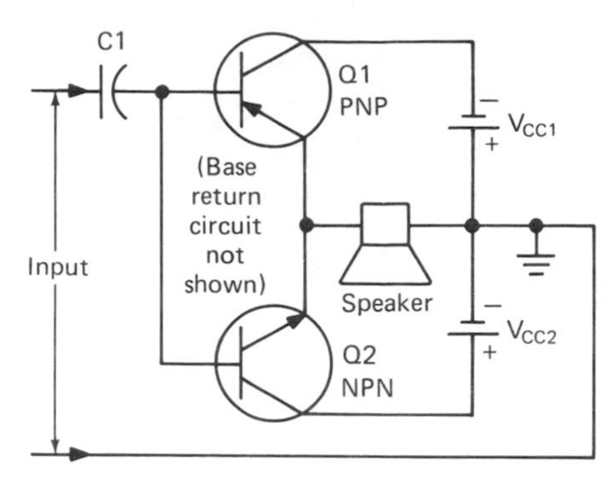

Skeleton circuit diagram for the basic *complementary-symmetry* arrangement. A PNP and an NPN transistor are operated in the emitter-follower (common collector) mode with zero bias (class B).

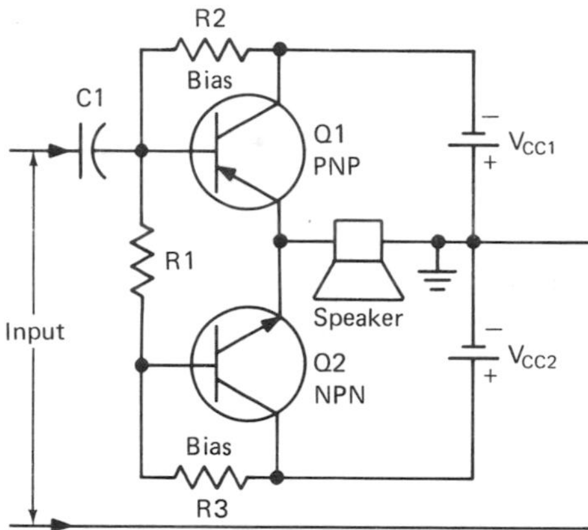

A *complementary-symmetry* configuration in which the transistors are operated with a small forward bias (class AB) to minimize crossover distortion. The PNP and NPN transistors function in the emitter-follower mode.

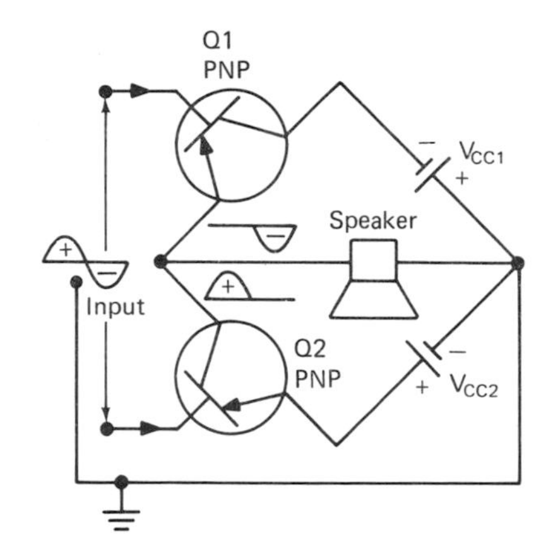

Skeleton circuit diagram for the basic *bridge arrangement* of a *complementary-symmetry* output stage. Q1 operates in the emitter-follower mode, whereas Q2 operates in the common-emitter mode.

Figure 3-9 Some basic types of audio power amplifiers.

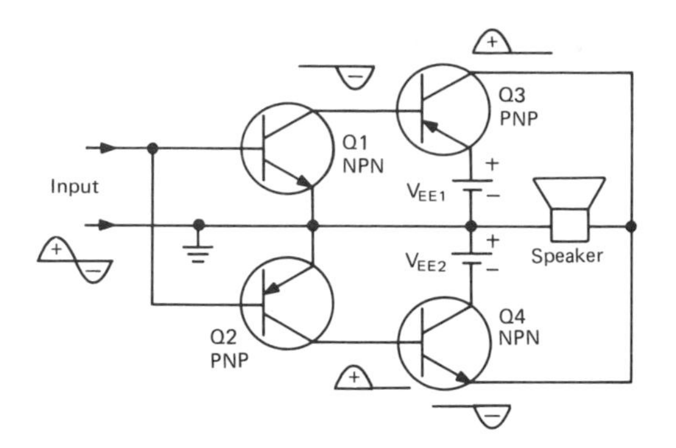

Common-emitter version of the direct-coupled complementary-symmetry amplifier. The transistors operate in class B. Class AB operation may be employed to minimize crossover distortion.

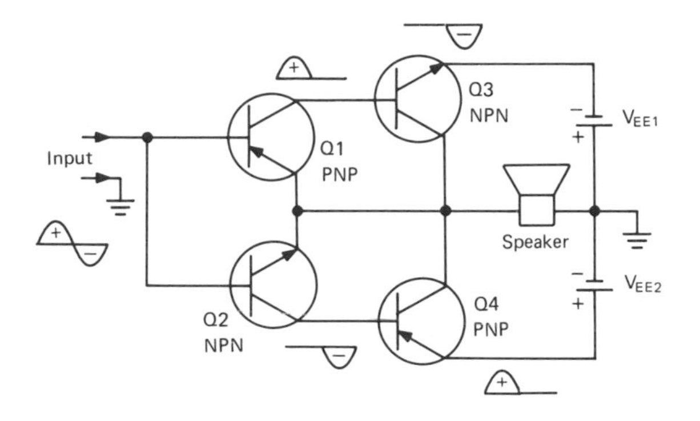

Common-collector version of the direct-coupled complementary symmetry amplifiers. The stage gain is less than in the common-emitter arrangement, but the input resistance is greater.

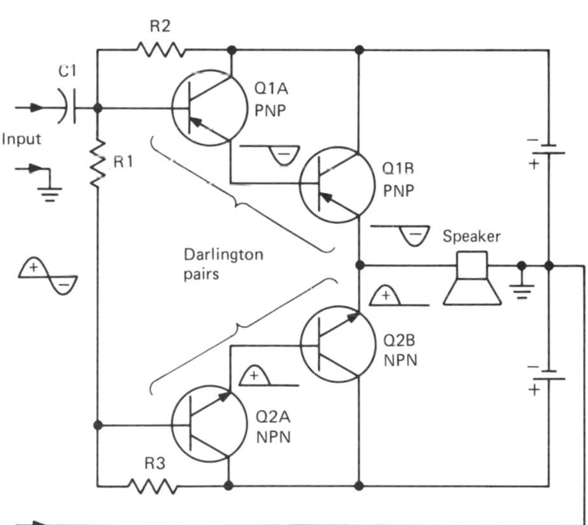

A complementary-symmetry *compound-connected* configuration. Q1A/Q1B and Q2A/Q2B operate as compound transistors, also called Darlington pairs, double emitter followers, or Beta multipliers. Each of the compound transistors functions as an emitter follower.

Figure 3-10 Basic direct-coupled complementary-symmetry stages.

Distortion Case History

Observe the complementary-symmetry stage exemplified in Figure 3-11. The trouble symptom under analysis was distortion—this distortion was most objectionable at low volume levels, and was less severe at higher volume levels. DC voltage measurements at the transistor terminals turned up the following facts:

1. The base-emitter voltage on Q205 would normally be in the order of 0.6V, but measured only a small fraction of a volt. (It was evident that the distortion was bias-related.)
2. When the bias voltage was monitored with a program signal inputted, it was observed that the bias voltage varied with the signal level. At times, the bias-voltage polarity would reverse.
3. A check of the voltage drop across bias diode D201 showed zero volts—it was apparent that the diode was shorted. Replacement of the diode restored the stage to normal operation.

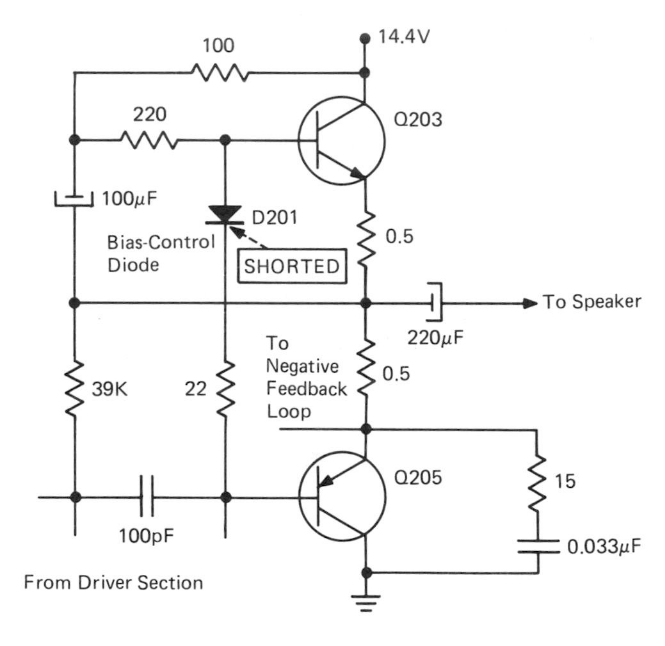

Note: This is a low-power amplifier, because it is rated for 0.9 watt power output. (High-power amplifiers are rated for more than 1 watt power output.)

Figure 3-11 Case history of distortion caused by a shorted diode.

NOISE LOCALIZATION

All audio amplifiers generate noise; in normal operation, the noise level goes unnoticed. An abnormal noise level is generally tracked down to a transistor with collector-junction leakage. However, resistors can also become noisy, as can diodes, leaky capacitors, or poor connections.

A noise voltage generated in the early stages (such as in the volume control) of an amplifier is comparatively troublesome, because the noise level is stepped up by subsequent stages. Of course, a large noise voltage generated in an output transistor can be just as apparent as a small noise voltage generated in an input transistor.

Noise sources are usually localized to best advantage as shown in Figure 3-12. A fixed capacitor is shunted across suspected devices or components, while monitoring the noise level from the speaker. When the capacitor is shunted across a noisy resistor, for example, the amplitude of the noise voltage is greatly reduced and the noise output from the speaker drops accordingly.

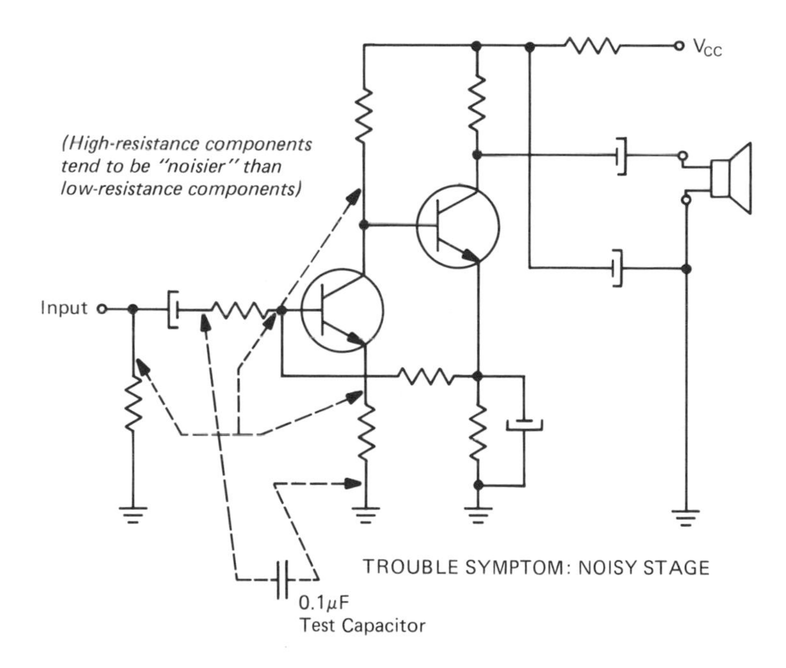

Figure 3-12 Bypass capacitor is used to localize a noise source.

Experiment

This experiment demonstrates self-oscillation from positive feedback, with the phono preamplifier constructed in the first chapter. Proceed as follows:

1. Connect a small speaker in series with a 0.05 μF capacitor; then connect the capacitor to the amplifier output terminal and connect the speaker to ground. (This arrangement avoids excessive loading of the preamplifier output circuit).
2. Connect a 50-pF capacitor from the amplifier's output terminal to its input terminal, and check for a tone output from the speaker.

In normal operation, an audio tone of approximately 1.5 kHz will be heard from the speaker. If a larger feedback capacitor, such as 0.002 μF is used, a much lower frequency audio tone will be heard from the speaker.

LOOK-AHEAD TROUBLESHOOTING

"Look-ahead" troubleshooting consists in noting down key voltage and resistance values for an amplifier before the need for service data arises. In other words, if you have an amplifier for which service data is not available, it is good foresight to take a little time and to jot down its key voltage and resistance measurements for possible future reference.

"IMPOSSIBLE" CASE HISTORY

Experienced troubleshooters know that tough-dog problems can result from "impossible" circumstances. As an illustration, a microphone amplifier was brought in for repair. It was completely dead. The amplifier was powered from a 9V battery that operated without a snap clip—the battery was merely inserted into a compartment. Because nearly all amplifiers of this type are designed with a clip connector and flexible leads, the technician was sufficiently curious that he or she examined the battery compartment terminals.

Surprisingly, it was discovered that the battery could make contact to the compartment terminals, regardless of the polarity with which the battery was inserted. The bottom line was that the customer had inserted a new battery in reverse polarity. Fortunately, the amplifier circuitry was not damaged, and the unit operated normally when the battery was correctly inserted.

4

Troubleshooting Radio Receivers Without Service Data

*Sizing Up the Job * Quick Checks * Capture Effect * Stage Identification in a Radio Receiver * Resonance Probe * High-Impedance Tuned Signal-Tracing Probe * Basic Comparison Tests * Identification of FM Transformers in FM/AM Receiver * Determination of Stage Sequence * Incidental Frequency Modulation * Addition or Subtraction of DC Voltages * Experiment*

SIZING UP THE JOB[1]

When you are troubleshooting a radio receiver, preliminary analysis is basically the same as for sizing up a malfunctioning audio system. Procedure is as follows:

1. Be a good listener and encourage your customer to talk. You can save valuable time, for example, if you know that the radio would operate normally only when turned upside down, before it went completely dead, or that it made a loud humming sound before it stopped operating.
2. If the radio is battery operated, check the battery voltage under load. It can happen that a short-circuited power branch will immediately ruin a new battery.[2]
3. Don't overlook the obvious: Are the battery leads intact? Do they make good connection? Is the speaker warped or otherwise damaged? Are its pigtail leads intact? Is the circuit board cracked? Has the board been corroded by a leaky battery? Do the controls have a normal feel?

---

[1]See Chart 4-1.
[2]See Chart 4-2.

CHART 4-1

Widely Used AM Radio Transistor Arrangements

These are some basic AM radio receiver arrangements to keep in mind when troubleshooting without service data.

STANDARD SIX-TRANSISTOR ARRANGEMENT

ALTERNATE SIX-TRANSISTOR ARRANGEMENT

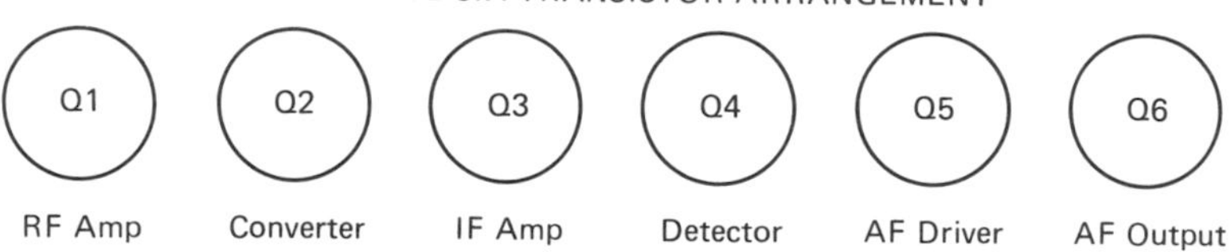

SIX-TRANSISTOR ARRANGEMENT WITH PUSH-PULL AUDIO OUTPUT

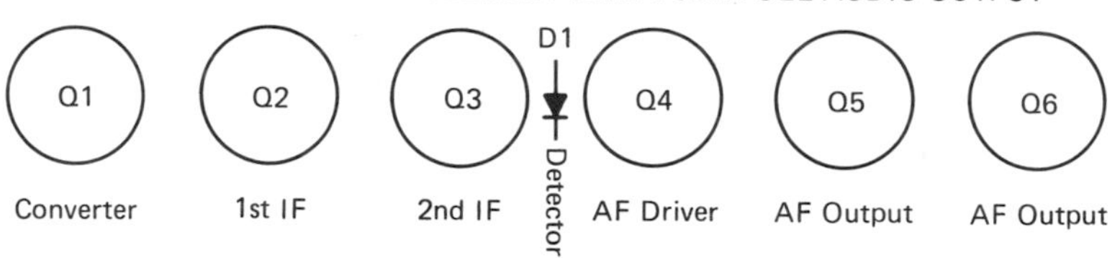

"ALL-AMERICAN-FIVE" RECEIVER ARRANGEMENT

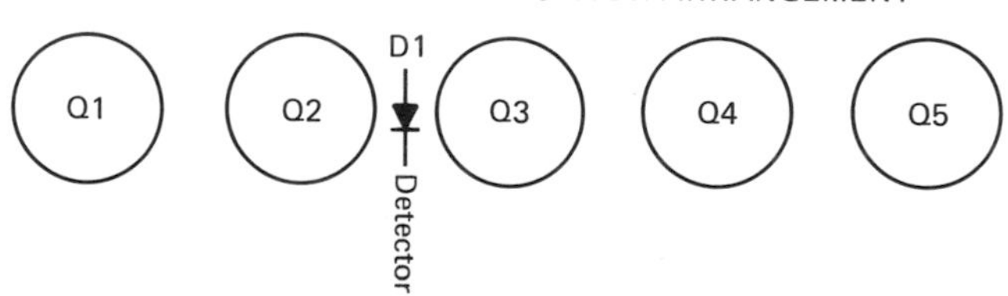

Radio circuit boards sometimes have parts numbers marked beside the devices and components, starting with "1" for input parts, and increasing sequentially through the network, with the highest parts numbers marked beside the output devices and components.

The standard six-transistor arrangement and the alternate six-transistor arrangement employ transistor detectors. A transistor detector is essentially a demodulator and audio amplifier.

CHART 4-2

Impedance Check at Battery Clip Terminals

Tough-dog troubleshooting problems can be caused by open capacitors (and capacitors with a poor power factor) associated with the V_{CC} line.

Such defective capacitors can often be caught at the outset by making a comparison check of the impedance Z at the battery clip terminals.

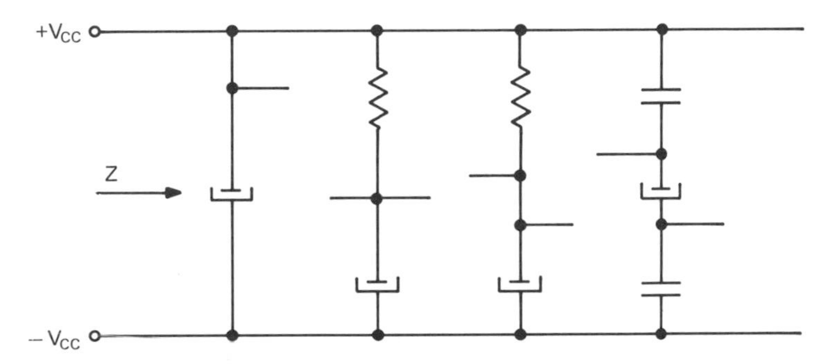

An impedance comparison check of a good receiver and a bad receiver is made with an audio generator, a DVM, and a resistor, as shown below.

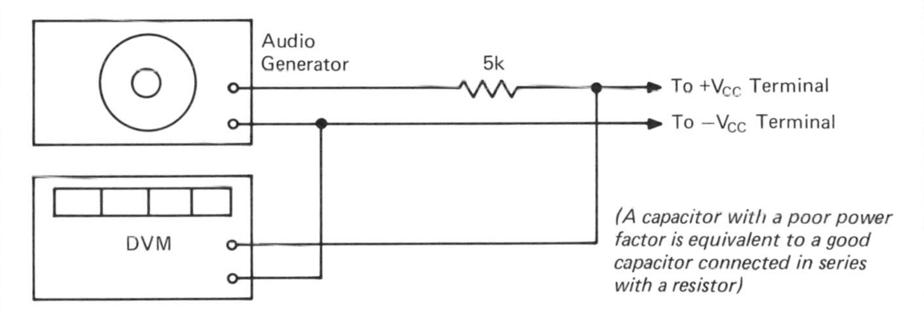

(A capacitor with a poor power factor is equivalent to a good capacitor connected in series with a resistor)

The comparison impedance check is made by disconnecting the battery from the receiver, and correcting the test leads of the checker to the V_{CC} terminals. Operate the audio generator at a low frequency, such as 60 Hz. Advance the output voltage from the generator to obtain an adequate AC-voltage indication on the DVM. Then, repeat the test on the other receiver, and compare the two readings.

Example: A typical pocket radio receiver in normal operating condition produces a reading of 20 MV on the DVM when the

CHART 4-2 CONTINUED

foregoing impedance test is made. If the reading were significantly excessive, such as 100 MV, or 500 MV on a bad receiver, the troubleshooter would look for one or more capacitors in the V_{CC} circuit with defects such as poor power factor, open circuit, or loss of capacitance conditions.

As noted previously, a quick check of a suspected open capacitor (or loss of capacitance, or poor power factor) can be made by temporarily bridging a good capacitor across the terminals of the suspect capacitor. Then, if normal operation resumes, the troubleshooter has localized the fault.

Useful Old and New Quick Checks

It is helpful to briefly recap some standard quick checks, along with informative follow-up tests for AM radio receivers.

1. If you are checking a battery radio, place your ear near the speaker; turn the power switch on and off. You will hear a click if the audio section is operative, if the speaker is workable, and if the earphone jack is not defective. If you do not hear a click, follow up by plugging in an earphone and repeating the click test.
2. In case the receiver passes the click test, turn up the volume control and listen for any noise (hissing) output. A reasonable amount of output noise throws suspicion on the receiver input circuitry (the local oscillator may have dropped out). Very low noise output directs attention to the detector circuit or the audio driver stage.
3. Follow up a very-low-noise condition with an amplifier substitution test. Locate the detector (it will be a diode), and feed the detector output into a miniamplifier with a built-in speaker, such as the Archer (Radio Shack) 277-1008. You will now hear a normal noise level. If the signal sections are workable, you will also hear stations as the receiver tuning control is turned.
4. If you do not hear station signals in the foregoing test, follow up by connecting a 1N34A diode in series with the miniamplifier input lead, and feeding the detector input signal into

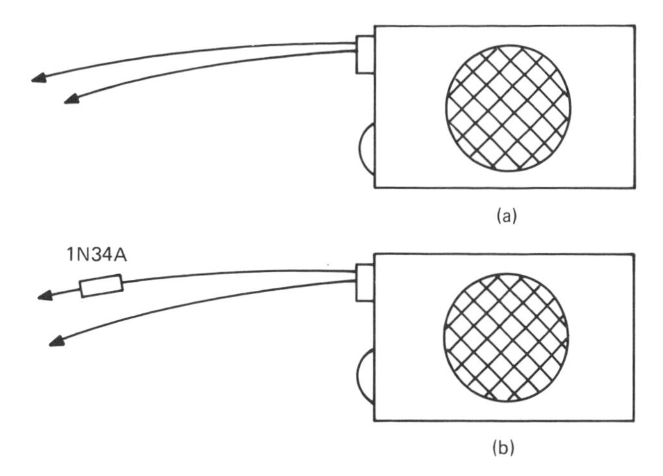

Technical Note: A miniamp/speaker functions as a "patch" section for quick checking the audio and speaker circuitry in a radio receiver. When a diode is included in the test lead, the tester also functions as a "patch" for the detector in the radio receiver.

Figure 4-1 Two important quick checkers. (a) Miniamp/speaker with test leads; (b) miniamp/speaker with diode in series with the "hot" test lead.

the 1N34A diode. If you now hear station signals as the receiver tuning control is turned, the detector diode in the receiver is defective. (See Figure 4-1.)

5. On the other hand, if you do not receive station signals using the substitute detector and amplifier, the trouble will be found in the RF (converter), IF, or AGC sections. Follow-up troubleshooting requires signal-section identification tests, as explained subsequently.
6. There is still one more standard quick check that you will usually wish to make at this point. Lack of station-signal reception is sometimes caused by local-oscillator drop-out. In such a case, reception will be restored if the RF output from a signal generator is coupled into the converter circuit, or the antenna coil. (The generator must be tuned to the appropriate oscillator frequency.)

WHAT HAPPENS IN THE CASE OF AN FM RECEIVER

If the radio receiver has an FM function, repeat the foregoing quick checks and compare the results with findings on its AM function. However, note the following technical points:

1. Most FM/AM radios employ more transistors on the FM function than on the AM function. In turn, the normal current drain with the volume control turned down is somewhat higher. For example, a radio that draws 9 mA on its AM function draws 11.5 mA on its FM function. (Connect a milliammeter in series with the battery.)
2. To check for a dead local oscillator, place the defective FM receiver near a normally operating FM receiver. In this case, the technician is not listening for "squeals"—he or she is listening for "mutes."

   Example: When the operating FM receiver is tuned to a station at 107 MHz, and the tuning dial of the dead receiver is varied from 88 to 98 MHz, a "dead silence" at 96.3 MHz (107−10.7 MHz) indicates that the local oscillator in the dead receiver is workable.

Capture Effect

The foregoing quick check of local-oscillator operation in an FM receiver is based on the *capture effect.* In other words, when two FM carriers are present at the same frequency, the weaker signal will be completely suppressed and the stronger signal will capture the FM detector. In this quick check, the weaker FM signal is the station to which the operative FM receiver is tuned. The local-oscillator radiation from the supposedly dead FM receiver is a stronger FM signal, because the receivers are side-by-side. Inasmuch as the radiated local-oscillator signal is unmodulated, capture results in silencing of the operative receiver.

You will observe that when the supposedly dead FM receiver is moved away from the operative receiver, the capture effect ceases and the operative receiver reproduces the station signal to which it is tuned.

Then, as you move the supposedly dead FM receiver still farther away from the operative receiver, the capture effect again occurs, and the operative receiver is again silenced. This test result occurs because of standing waves in the room. The local-oscillator radiation in this example has a frequency of 107 MHz, and the peaks of the standing waves are spaced 4.6 feet apart.

If a receiver is operative on its AM function, but is inoperative on its FM function, the trouble area is narrowed down considerably. (A receiver can be operative on its FM function, but inoperative on its AM function.) It follows that the troubleshooter needs to have a general understanding of the common sections and the separate sections in FM/AM receiver configurations.

STAGE IDENTIFICATION IN A RADIO RECEIVER

One of the requirements after quick checks are completed is to identify the RF, oscillator-mixer, IF, detector, and AF transistors. *As soon as stage identification is made, the transistor can be "color-mapped"* as previously explained for audio-amplifier troubleshooting.

Stage identification is easily and quickly accomplished with the aid of an oscilloscope (unless the stage is completely dead). As shown in Figure 4-2, signal frequencies are easily identified on a scope screen with respect to the sweep speed.

The receiver under test should be driven by a 1-MHz unmodulated signal from an AM signal generator. In turn, when the receiver is tuned to 1 MHz, an RF signal will be displayed as a sine wave with a period of 1 microsecond, an IF signal will be displayed as a sine wave with a period of 2.2 microseconds, and the local-oscillator signal will be displayed as a sine wave with a period of 0.69 microsecond. Note that the local-oscillator signal is displayed in the absence of an RF input signal.

To identify audio-frequency transistors, the RF input signal should be amplitude-modulated at approximately 30 percent. (The modulating frequency is usually 1 kHz.) The scope sweep speed should be set to 1 millisecond for full-screen deflection. In turn, when an audio-frequency transistor terminal is contacted with the scope's vertical-input lead, one cycle of a sine wave is displayed on the screen.

To obtain adequate vertical deflection when checking low-level circuits, a sensitive scope should be used. As a general rule, a scope with considerable bandwidth has lower vertical sensitivity. Thus, a scope with 2-MHz bandwidth might have a vertical sensitivity of 20 millivolts per centimeter, whereas a scope with 15 MHz bandwidth might have a vertical sensitivity of 0.1 volt (100 millivolts) per centimeter.

If your oscilloscope has insufficient sensitivity for checking low-

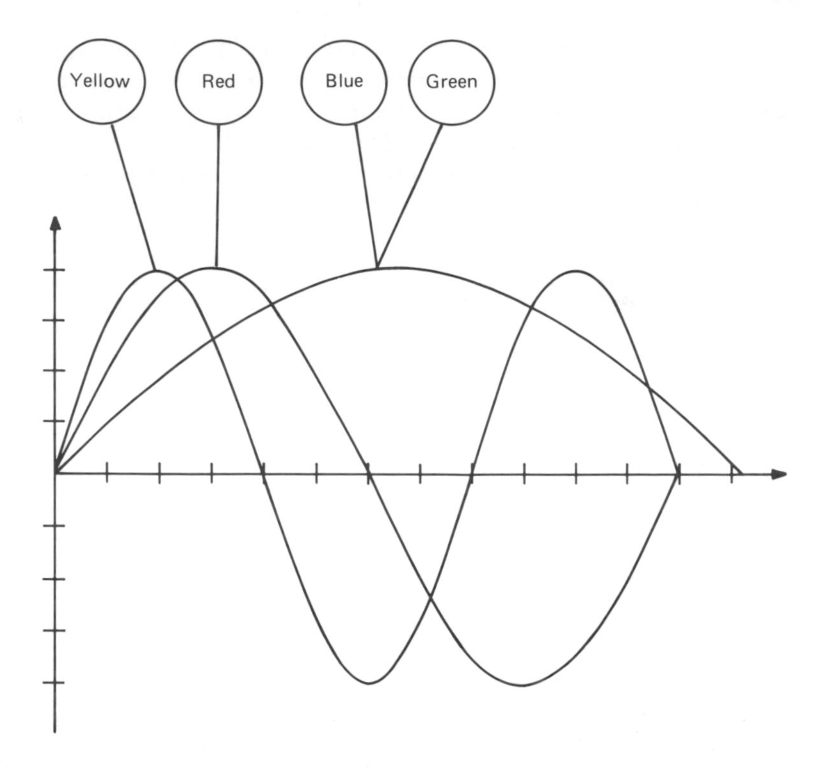

A1-MHz RF wave, a 1.455-MHz oscillator wave, and a 455-kHz IF wave are displayed in this manner on an oscilloscope screen when the sweep speed is the same for each wave. Thus, the "red" transistor is operating at 1 MHz, the tested terminal on the "yellow" transistor is operating at 1.455 MHz, and the "green" and "blue" transistors are operating at 455 kHz. This is just another way of saying that "red" is an RF circuit, that "yellow" is a local-oscillator circuit, and that "blue" and "green" are IF circuits.

Figure 4-2 Stage identification in an AM radio receiver.

level circuits in radio receivers, it can be used with a preamplifier to increase its gain by 100 times, for example. This topic is explained in greater detail subsequently.

STAGE IDENTIFICATION IN DEAD CIRCUITRY (RESONANCE PROBE)

When troubleshooting without service data, stages can be easily identified in dead circuitry by means of a resonance probe, as shown in Figure 4-3.

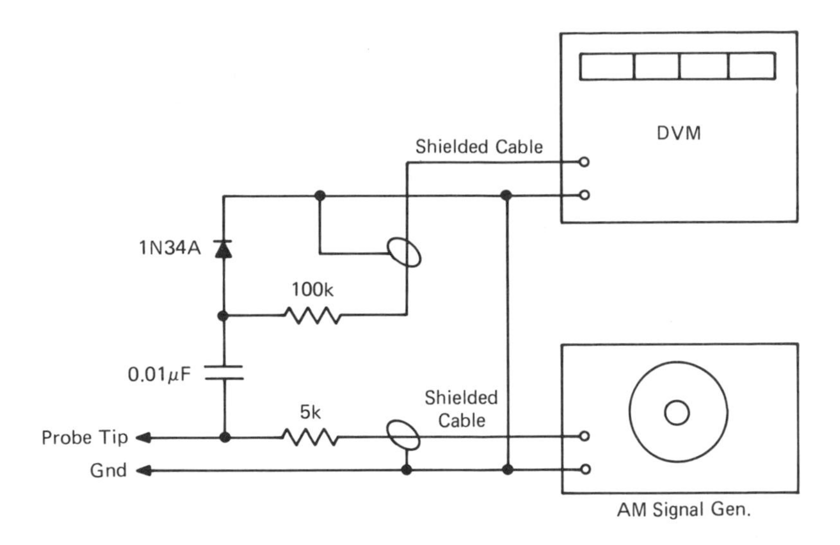

A resonance probe operates on the basis of RF (or IF) voltage variation vs. frequency. When the probe-tip and ground clips are connected across a tuned coil (or transformer), the resonance probe "sees" an impedance that is frequency-dependent. In turn, the DVM indicates a very low voltage when the generator is set off-frequency; on the other hand, the DVM indicates a comparatively high voltage when the generator is set to the resonant frequency of the tuned coil (or transformer) under test. Thereby, the coil is identified as an RF coil, a local-oscillator coil, or an IF coil. Note that the resonance probe also develops more or less voltage magnification. In other words, when the probe-tip and ground clips are connected across a coil and the generator is adjusted to resonance, the DVM typically indicates a voltage substantially greater than the open-circuit voltage when the probe tip is disconnected from the coil under test.

Note that the resonance probe has input capacitance and detunes the coil under test to some extent. In other words, the coil resonates to a somewhat lower frequency than in usual operation, when the resonance probe is connected across the coil.

Some service AM signal generators have comparatively low-level output that provides zero indication on a DVM with a peak-reading probe. In this application, *a generator is required which has sufficiently high output to provide an indication of 50 or 100 millivolts on the DVM.*

Figure 4-3 Resonance probe is used for stage identification in dead receivers.

A resonance probe shows whether a coil (or transformer) is an RF, local-oscillator, or IF coil, on the basis of frequency indication (unless the coil under test is open or shorted). The resonance probe is basically a peak-indicating probe energized by an RF (or IF) signal voltage through an isolating resistor. From a functional viewpoint, a resonance probe can be regarded as a "reverse dip meter." Because the peak test voltage is less than 0.5 volt, the resonance probe does not turn on transistors or diodes that may be connected to the coil under test—the resonance probe "sees" only the coil.

This is just another way of saying that the resonance probe utilizes a germanium diode, whereas the radio receiver under test employs silicon devices. In turn, the resonance probe is activated by peak voltages above 0.2 volt, but the receiver's silicon devices are not activated until the peak voltage exceeds 0.5 volt.

EXAMPLE 1

DVM reading before resonance probe is connected across coil in the receiver under test: 48 mV.

DVM reading when resonance probe is connected across coil in the receiver under test: 13 mV.

DVM reading when generator is tuned to 748 kHz: 72 mV.[3]

Therefore, the coil under test is an RF coil.

EXAMPLE 2

DVM reading before resonance probe is connected across coil in the receiver under test: 57 mV.

DVM reading when resonance probe is connected across coil in the receiver under test: 1.5 mV.

DVM reading when generator is tuned to 450 kHz: 47 mV.[4]

Therefore, the coil under test is an IF coil. (Note that the generator tuning is quite critical when an IF coil is being checked.)

EXAMPLE 3

DVM reading before resonance probe is connected across coil in the receiver under test: 33 mV.

DVM reading when probe is connected across coil in receiver under test: 3 mV.

DVM reading when generator is tuned to 1.95 MHz: 23 mV.

---

[3]This test result represents a voltage magnification of 24 mV (an effective Q value of 1.5).

[4]This test result represents a voltage magnification of less than unity, or −10 mV (an effective Q value of 0.8).

Therefore, the coil under test is a local-oscillator coil. (Note that the generator tuning is very critical when a local-oscillator coil is being checked.)

HIGH-IMPEDANCE TUNED SIGNAL-TRACING PROBE

When you are troubleshooting radio receivers without service data, live stages can be identified with an oscilloscope as explained above. A live stage can also be identified with a tuned signal-tracing probe. (A tuned probe responds to an RF signal, for example, but not to an IF signal, nor to a local-oscillator signal.) Tuned signal-tracing probes include one or more transistors to provide sufficient gain for tests in low-level radio circuits.

Since junction-type transistors have comparatively low-input impedance, and impose appreciable circuit loading when employed in tuned signal-tracing probes, the troubleshooter may prefer the tuned high-impedance signal-tracing probe shown in Figure 4-4. This

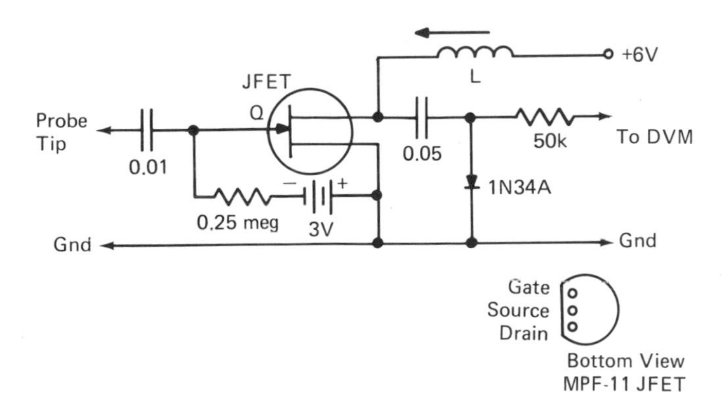

Note: Transistor Q may be a type MPF-11 N-channel JFET, or equivalent. L is an AM broadcast-band inductor. The active signal-tracing probe consists of an RF-amplifier stage driving a peak detector.

When the probe is directly applied across a high-Q tuned coil with the same resonant frequency as L, instability could occur (the DVM could indicate a signal voltage when no signal is present). In such a case, connect an isolating resistor of suitable value in series with the probe-tip (the isolating resistor should have a sufficiently high value that the DVM indicates zero signal when no signal is present).

Figure 4-4 Sensitive high-Z high-frequency rectifier probe: Active AM broadcast high-impedance RF signal-tracing probe.

probe employs a junction field-effect transistor (JFET). Since it is a tuned signal-tracing probe, it can be used to identify a live stage (to determine whether it is an RF, local-oscillator, or IF stage). It provides substantial gain and facilitates tests in low-level (or weak) stages. Inasmuch as the probe responds only to the frequency of its tuned drain coil, it must be used in conjunction with a signal generator set to the probe's frequency. A tuned-IF signal-tracing probe, however, has 455 kHz response, and is independent of the signal frequency in the RF section of the receiver.

The active JFET signal-tracing probe shown in Figure 4-4 has a voltage gain of approximately 35 times. Its input impedance is high, and its loading effect on the circuit under test is less than that of an active probe of the bipolar transistor type. It is also comparatively stable; in other words, it is less likely to exhibit a false RF output when connected across a high-Q coil tuned to the same frequency as L. If instability should be encountered, a comparatively small value of isolating resistance connected in series with the probe tip will stabilize the arrangement.

BASIC COMPARISON TESTS

When troubleshooting, the technician has a considerable advantage if he or she can locate a similar radio in normal operating condition. Two fundamental types of comparison tests may then be made:

1. With the receivers turned off, comparative resistance measurements can be made from corresponding test points to ground. These resistance measurements should be made with a low-power ohmmeter. Substantially different resistance values at corresponding test points indicate a fault in the associated circuit.
2. With the receivers turned on, comparative DC voltage measurements can be made from corresponding test points to ground. Substantially different voltage values at corresponding test points indicate a fault in the associated circuit.

IDENTIFICATION OF FM TRANSFORMERS IN FM/AM RECEIVER

When troubleshooting FM/AM receivers, the technician sometimes needs to identify the FM IF transformers and the AM IF transformers in a dead system. This can be easily done with the resonance probe depicted in Figure 4-3. There are two general situations encountered:

1. An FM transformer operates separately from an AM transformer.
2. An FM transformer operates in series with an AM transformer, as shown in figure 4-5. (See also Figures 4-6 and 4-7.)

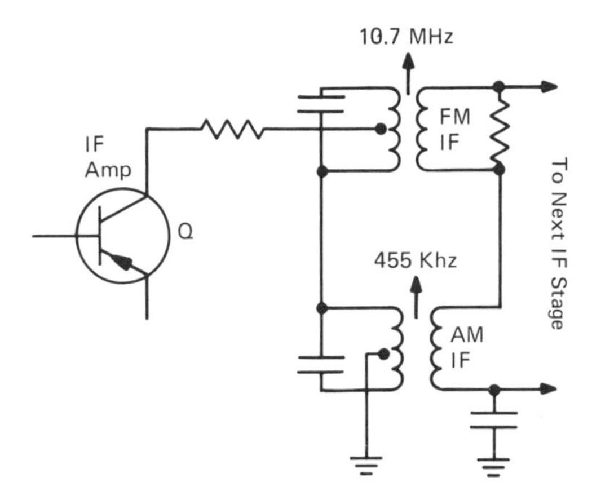

The FM and AM IF transformers operate with their primary windings connected in series, and with their secondary windings connected in series. When Q outputs a 455-kHz signal, the 10.7-MHz windings look like virtual short-circuits. On the other hand, when Q outputs a 10.7-MHz signal, the 455-kHz windings have very low impedance, and look like virtual short-circuits.

When a resonance probe is applied between the collector terminal of Q and ground, a resonant voltage rise will normally occur at 455 kHz, and at 10.7 MHz. The two IF transformers can be identified as follows:

1. When the primary winding of the 10.7-MHz transformer is short-circuited with a jumper, only the 455-kHz resonant voltage rise normally occurs.
2. When the primary of the 455-kHz transformer is short-circuited with a jumper, only the 10.7-MHz resonant voltage rise normally occurs.

Note that this type of IF amplifier normally has V_{CC} voltage present whether the function switch is set for AM reception, or for FM reception. A stage that is used for FM reception only will normally have V_{CC} voltage present only when the function switch is set to its FM position. Similarly, a stage that is used for AM reception only will normally have V_{CC} voltage present only when the function switch is set to its AM position.

Figure 4-5 Basic FM/AM IF transformer configuration.

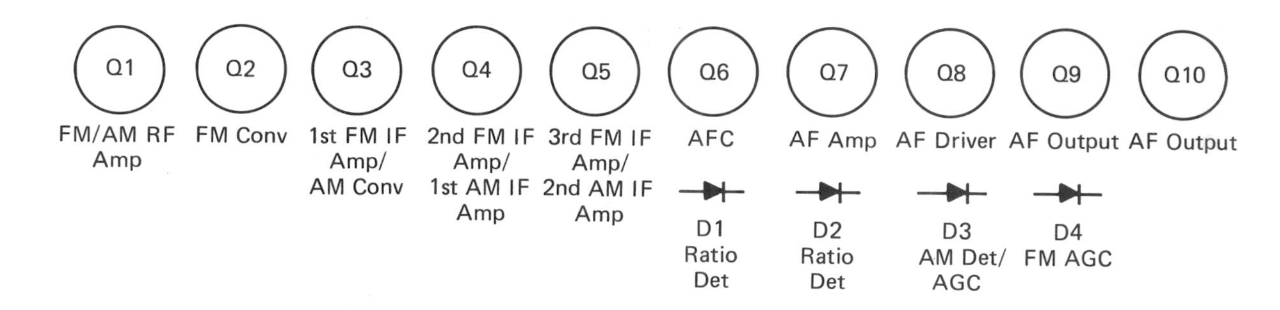

Note that an FM/AM receiver generally employs more transistors on its FM function than on its AM function. In this example, the receiver operates as an 8-transistor AM configuration, and as a 10-transistor FM configuration. One diode is utilized on the AM function, and three other diodes are used on the FM function. (Older designs of AM/FM receivers included an additional transistor for the FM local-oscillator circuit; the more recent designs combine a local-oscillator circuit and a mixer circuit into a single FM converter circuit.) Refer to Figure 4-7 for a widely used ratio-detector configuration.

Noisy and/or distorted sound reproduction on the FM function is often caused by a fault in the ratio-detector circuit. Check the ratio-detector diodes and the capacitors in the circuit.

Figure 4-6 Popular FM/AM transistor complement: Transistor lineup for a 10-transistor FM/AM receiver.

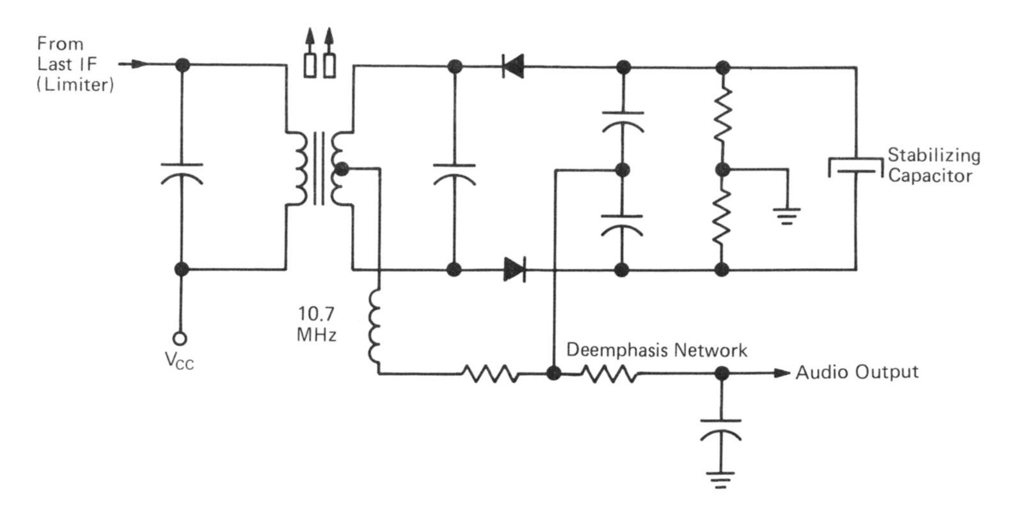

Distortionless FM detection and optimum rejection of AM requires proper alignment, matched diodes, and a stabilizing capacitor with good insulation resistance, full rated capacitance, and a good power factor. Trouble symptoms can also be caused by leakage or opens in the smaller fixed capacitors, or by off-value resistors. Although the ratio-detector transformer can develop faults, this is an infrequent occurrence.

A ratio detector operates like a discriminator, and has a *capture effect.* Normally, a weaker interfering signal will be captured by the ratio detector and completely suppressed, so that the listener is unaware of its presence.

Normally, a *quieting effect* is provided by the ratio detector, whereby the noise output between channels is suppressed when an unmodulated or modulated carrier is tuned in. The quieting sensitivity of an FM receiver is the minimum input signal level that will give a specified signal/noise ratio at the detector output.

Figure 4-7 A widely used ratio-detector configuration. Ratio-detector circuit normally rejects AM.

In either case, a transistor will be mounted near the transformer(s). When a resonance probe is applied between the collector of the transistor and ground (common), two types of response will normally be observed:

1. A resonant voltage rise occurs at one frequency (either at 455 kHz or at 10.7 MHz).
2. A resonant voltage rise occurs at two frequencies (455 kHz and 10.7 MHz).

If a voltage rise occurs at 455 kHz, the troubleshooter knows that the transistor is connected to the nearby AM IF transformer.

If a voltage rise occurs at 10.7 MHz, the troubleshooter knows that the transistor is connected to the nearby FM IF transformer.

If a voltage rise occurs at both 455 kHz and at 10.7 MHz, the troubleshooter knows that the transistor is connected to the nearby IF transformers—one of these will be an AM transformer, and the other will be an FM transformer.

Although FM and AM transformers look the same, they can easily be identified by a short-circuit test. In other words, when the troubleshooter short-circuits the terminals of one transformer with a jumper, either the 455-kHz or the 10.7-MHz voltage rise will be killed. If the 455-kHz voltage rise is killed, the troubleshooter knows that the jumper is connected to an AM transformer—and vice versa.

PRACTICAL EXAMPLE:

DVM reading before resonance probe is connected from collector to ground in Figure 4-5: 129 mV.

DVM reading after resonance probe is connected from collector to ground in Figure 4-5: 10 mV.

DVM reading when generator is tuned to 455 kHz: 115 mV.

HIGH-BAND CHECK

DVM reading before resonance probe is connected from collector to ground in Figure 4-5: 25 mV.

DVM reading after resonance probe is connected from collector to ground in Figure 4-5: 7 mV.

DVM reading when generator is tuned to 10.7 MHz: 19 mV.

The foregoing results show that an AM transformer is connected in series with an FM transformer from the transistor's collector terminal to ground. Note that both transformers will be mounted in the vicinity of the transistor. Note also that the FM transformer can be distinguished from the AM transformer by short-circuit jumpering, as explained above.

DETERMINATION OF STAGE SEQUENCE

Stage sequence in a dead system can be determined by signal injection/tracing tests. In other words, if a test signal is injected at the collector of a "red" transistor, and a signal-tracing probe shows that the signal appears at the base of a "blue" transistor, the troubleshooter concludes that the "blue" stage follows the "red" stage. (Use an in-circuit transistor tester to check transistor basing.)

This method is workable whether a transistor is leaky or open, but is not workable if a tuned transformer is short-circuited, for example, or if there is an open circuit in the signal path. However, determination of stage sequence can then be made on the basis of available "landmarks" and visual tracking of the PC conductors. Since circuit details may be comparatively involved, it is generally quicker and easier to use the signal injection/tracing method.

> ***Incidental Frequency Modulation (IFM):*** **When checking FM circuitry with an AM generator signal, the troubleshooter can become puzzled concerning the response of the FM detector to an amplitude-modulated signal. This response occurs because service AM generators usually have incidental frequency modulation, particularly when operated at higher percentages of modulation.**

ADDITION OR SUBTRACTION OF DC VOLTAGES

Occasionally the troubleshooter desires to monitor the sum or the difference of two voltages during troubleshooting procedures. This is easily done, as shown in Figure 4-8, for two voltages that have a common ground. The DC voltmeter is merely connected between the two test points, and the scale indication is the difference between the two voltages.

To monitor the sum of two voltages, the microampere function of the VOM is used. Each of the voltage sources is connected through a 1-megohm resistor to the hot terminal of the microammeter, and the common terminal of the microammeter is connected to the common ground of the voltages. In turn, the sum of the voltages for the two test points is equal to the microampere indication on the scale.

Experiment

This experiment employs an ArcherKit (Radio Shack) No. 28-4029. The kit comprises devices and components for construction of a 2-transistor 1-IC AM radio. Before starting assembly, note that the kit includes a bar antenna with antenna coil, an IF transformer color-coded yellow, an IF transformer color-coded black, and a converter transformer color-coded red. (See Figure 4-9.)

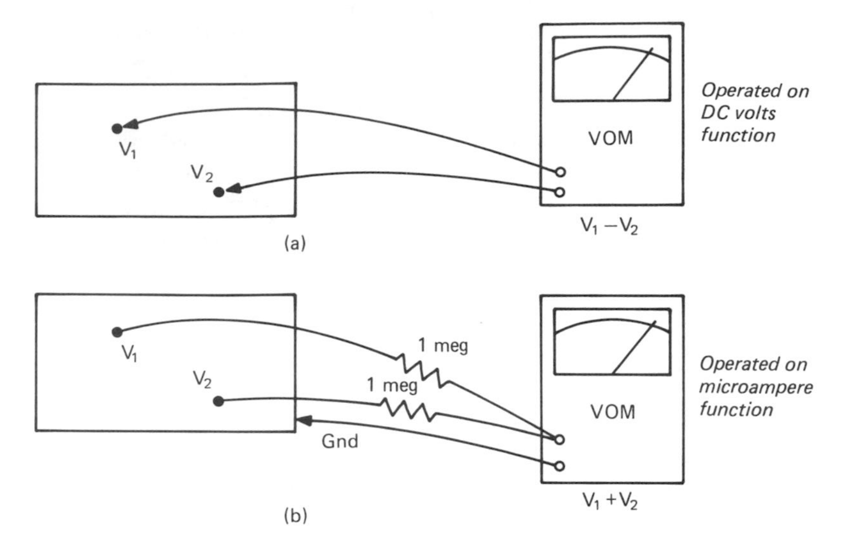

To subtract two voltages with respect to a common ground, the DC voltmeter is connected directly between the two sources.

To add two voltages with respect to a common ground, the hot terminal of the DC microammeter is connected via 1-megohm resistors to each of the sources, and the common terminal of the DC microammeter is connected to the common ground. The sum of the two voltages is equal to the microampere reading.

(The same method can be used to monitor the sum of three voltages, if desired.)

A reasonably sensitive VOM should be used to measure the sum of two or more voltages. For example, the Micronta (Radio Shack) 50,000 ohms/volt multitester is suitable.

Figure 4-8 Addition or subtraction of DC voltages. (a) Subtraction of V_1 and V_2; (b) addition of V_1 and V_2.

These inductive components will show the following responses when checked with a resonance probe:

1. Check the 3-terminal winding (primary) of the "black" IF transformer with the resonance probe. Observe that a voltage rise in the vicinity of 455 kHz is obtained when the probe is connected to any two terminals. Observe that a similar (although smaller) voltage rise is obtained when the probe is connected across the 2-terminal winding (secondary).
2. Repeat the foregoing checks on the "yellow" IF transformer.

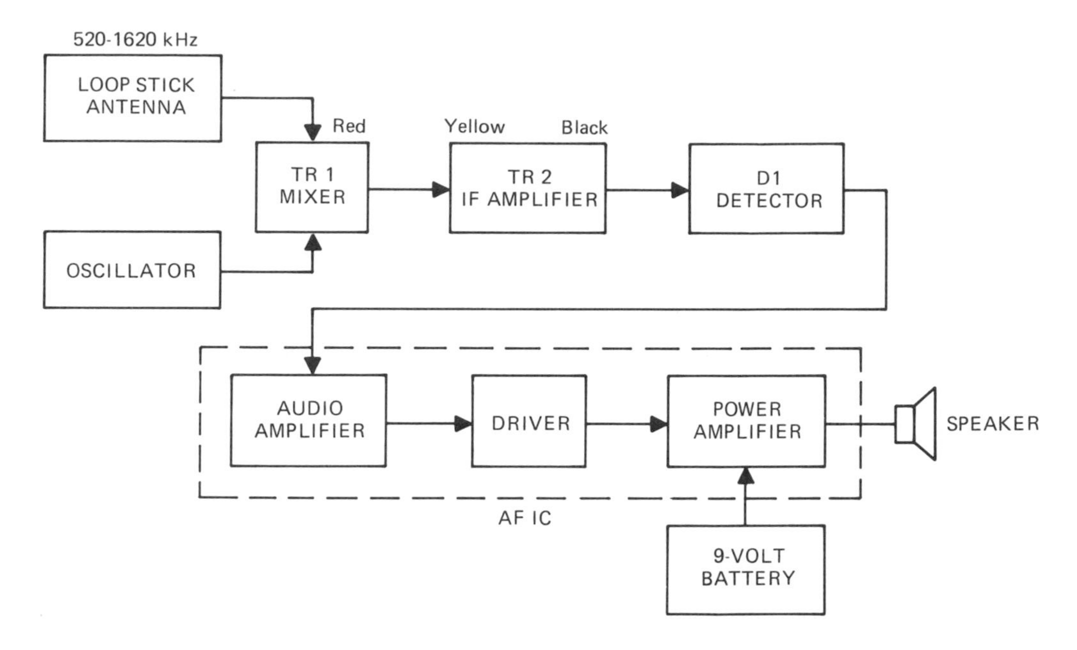

Note: The oscillator and mixer employ the same transistor, and this combined circuitry is called a converter stage. It operates with two tuned components: the loopstick antenna (bar antenna) and the oscillator coil (coded red). The IF amplifier operates with an input transformer (coded yellow) and an output transformer (coded black).

These tuned components are easily checked with a resonance probe.

Figure 4-9 Block diagram of 2-transistor 1-IC AM radio.

3. Check the 3-terminal winding (primary) of the "red" converter transformer. Observe that a voltage rise in the vicinity of 2.25 MHz is obtained when the probe is connected to any two terminals. Observe next that no discernible voltage rise is obtained when the probe is connected across the 2-terminal winding (secondary)—because the secondary is very loosely coupled to the primary.
4. Check the primary winding of the bar antenna coil with the resonance probe. Observe that a voltage rise in the vicinity of 1.5 MHz is obtained; note also that this is a high-Q winding, and that some voltage magnification is obtained. Observe next that no discernible voltage rise is obtained when the probe is connected across the secondary winding of the bar antenna coil—because this is a very small winding with little effective coupling to the primary.

Construction and further checkout of the AM radio receiver will be described in Chapter 5.

Note on Generator Harmonics: **When troubleshooting with resonance probes, remember that service AM generators usually have appreciable harmonic output. Thus, when the generator is set to 455 kHz, it will often have appreciable output at 910 kHz, and may have noticeable output at 1365 kHz. These spurious outputs can cause confusion in some test situations, if the troubleshooter overlooks their presence.**

5

Additional Radio Troubleshooting Techniques

*Comparative DC Voltage and Resistance Checks * Practical Considerations * Example of Troubleshooting by Comparative Voltage and Resistance Measurements * Reminder * Comparative DC Voltages With and Without Signal * Series vs. Shunt Detection * Note on Circuit Disturbance * Converter Voltages in Oscillatory and Non-Oscillatory States * All Voltages Negative * Oscillator Signal Substitution * Experiment*

COMPARATIVE DC VOLTAGE AND RESISTANCE CHECKS

An outstanding advantage of preliminary troubleshooting by comparative DC voltage and resistance checks is that the technician does not need to buzz out any of the receiver circuitry. Its chief limitation is that a similar and normally operating receiver may not be available. When comparison tests are feasible, the receivers should be tuned to the same frequency. Examples of comparative DC voltage data for a pair of similar receivers, with and without signal present, are as follows:

AM RADIO RECEIVER

Comparative DC voltage checks of converter transistors:

Collector voltages, 8.99 V vs. 8.84 V (a difference of 0.15 V).
Base voltages, 2.70 V vs. 2.50 V (a difference of 0.20 V).
Emitter voltages, 2.32 V vs. 2.10 V (a difference of 0.22 V).
(DC voltages are practically the same whether signal is present or absent.)

} Significantly different values would indicate a fault in one of the sections

Note that when comparative DC voltage measurements are made by connecting the DVM from collector-to-collector, from base-to-base, or from emitter-to-emitter between transistors in corresponding receivers, a jumper must be connected from the ground of one receiver to the ground of the other receiver, in order to complete the measuring circuit. (See also Chart 5-1.)

Comparative DC voltage checks of audio output transistors:

Collector voltages, 9.07 V vs. 9 V (a difference of 0.07 V). Base voltages, 5.19 V vs. 5.11 V (a difference of 0.08 V). Emitter voltages, 5.84 V vs. 5.73 V (a difference of 0.11 V). (DC voltages on base are slightly greater with signal present.)	Significantly different values would indicate a fault in one of the sections

Observe that the base-emitter bias voltage for the audio output transistors is approximately 0.64 V in the foregoing example, showing that the transistors are operated in class A. This is the reason that the DC base voltage shifts but slightly from no signal to maximum signal conditions.

Observe next that the base-emitter bias voltage for the converter transistors is approximately 0.4 V in the foregoing example—*seemingly an off-value bias voltage.* However, this is not so—inasmuch as this is a converter transistor, it is in an oscillatory condition while measuring its terminal voltages. The amplitude of oscillation is normally high enough to produce some rectified base-emitter current flow which opposes the effective bias voltage—in this example, from 0.6 volt to 0.4 volt.

Otherwise stated, the base of a converter transistor is normally driven alternately into cutoff and into conduction by the local-oscillator voltage; the base conducts in spurts on the positive peaks of the oscillator voltage. The base seems to be cut off all the time because a DC meter cannot follow the positive and negative swings of the oscillator AC voltage.

Comparative Resistance Checks

Comparative resistance-to-ground checks of converter transistors:

Low-power ohmmeter used; battery disconnected; jumper connected from $+V_{CC}$ to $-V_{CC}$ receiver terminals.

CHART 5-1

Practical Considerations in Preliminary Troubleshooting Procedures

Transistor terminals are sometimes accessible only from the solder side of the PC board. In other words, the base of the transistor may be in contact with the PC board, or the transistor may be so closely surrounded by other components or devices that test connections to the transistor pigtails are impractical.

In such a case, the transistor terminals will be accessible from the solder side of the PC board. Test connections to the solder pads can be made with sharp-pointed test prods. It is sometimes helpful to "tack" test pigtails to the solder pads with a pencil-type soldering iron.

After a test pigtail has been "tacked" to a solder pad, miniature clips can be easily connected, just as if tests were being made on the component side of the PC board. After a repair job has been completed, the test pigtails can be clipped off close to the solder pad with a pair of diagonal cutters.

Note that if a radio receiver has any audio output, FM transistors can be distinguished from AM transistors in preliminary trouble-shooting procedures by temporarily short-circuiting the base terminal of a transistor to its emitter terminal. In turn, if FM reception stops, the transistor under test operates in the FM channel. Conversely, if AM reception stops, the transistor under test operates in the AM channel.

An IF transistor often operates in both the FM channel and the AM channel. Of course, the AF transistors operate in both the FM and AM channels.

CHART 5-1 CONTINUED

In other situations, preliminary troubleshooting procedures involve receivers with up-front trouble. In such a case, the foregoing test procedure is not feasible. However, signal-injection tests can usually be made to distinguish FM transistors from AM transistors on the basis of response to 455 kHz and/or 10.7 MHz test frequencies.

Collector resistances, 6.6 ohms vs. 4.5 ohms, power switch on. "Crawl" toward infinity, power switch off.
Base resistances, 1.81 kilohms vs. 1.74 kilohms, power switch on or off.
Emitter resistances, 559 kilohms vs. 590 kilohms, power switch on. "Crawl" toward infinity, power switch off.
Collector resistances show a difference of 2.1 ohms.
Base resistances show a difference of 70 ohms.
Emitter resistances show a difference of 9 kilohms.

Significantly different values would indicate a fault in one of the sections, in this example.

Comparative resistance-to-ground checks of audio output transistors:

Low-power ohmmeter used; battery disconnected; jumper connected from $+V_{CC}$ to $-V_{CC}$ terminals.

Collector resistances, 2.9 ohms vs. 3.9 ohms, power switch on. Infinity, power switch off.
Base resistances, "crawl" toward infinity, power switch on or off.
Emitter resistances, 1.20 kilohms vs. 1.21 kilohms, power switch on. "Crawl" toward infinity, power switch off.
Collector resistances show a difference of 1 ohm.
Base resistances show no discernible "crawl" difference.
Emitter resistances show a difference of 10 ohms.

Significantly different values would indicate a fault in one of the sections, in this example.

EXAMPLE OF TROUBLESHOOTING BY COMPARATIVE VOLTAGE AND RESISTANCE MEASUREMENTS

In this example, an AM radio receiver had weak and noticeably distorted output. A similar receiver in normal operating condition was used for comparative voltage and resistance measurements. Three transistors and an integrated circuit were utilized, as depicted in Figure 5-1. (An educated guess would be that one transistor functions as a converter, that the two other transistors function as IF amplifiers, and that the integrated circuit functions as an audio amplifier.)

DC voltage measurements showed that the transistor terminal voltages were positive. A high-power ohmmeter check, with ground negative, showed that one of the transistor terminals had low forward resistance to ground. Accordingly, the transistors were identified as NPN types, and the base terminals were evident. (The emitter terminals were also evident, because they had either zero or near-zero voltage to ground.)

All of the collector DC voltages were somewhat subnormal in the defective receiver, although the base-emitter voltages made sense, with one exception. With respect to Q3 in Figure 5-1, the base voltage was high, although the collector voltage was low:

Good Receiver	Defective Receiver
Collector, 4.94 V	Collector, 3.34 V
Base, 0.726 V	Base, 0.76 V
Emitter, 0 V	Emitter, 0 V

These DC voltage measurements pointed to a fault in the Q3 circuit section. However, further tests were required to pinpont the fault. Accordingly, follow-up resistance checks were made. The test data are as follows:

(Measurements made with low-power ohmmeter, with battery disconnected from receiver)

Good Receiver (Q3)	Defective Receiver (Q3)
Resistance readings uncertain due to ohmmeter "crawl"	Resistance readings uncertain due to ohmmeter "crawl"

(Measurements made with low-power ohmmeter, with jumper connected from $+V_{CC}$ to $-V_{CC}$ terminals, and power switch on)

Good Receiver (Q3)	Defective Receiver (Q3)
Collector-Gnd, 490 ohms	Collector-Gnd, 490 ohms
Base-Gnd, 221 kilohms	Base-Gnd, 39 kilohms

The first set of resistance readings was uncertain because electrolytic capacitors in the circuit were causing prolonged meter

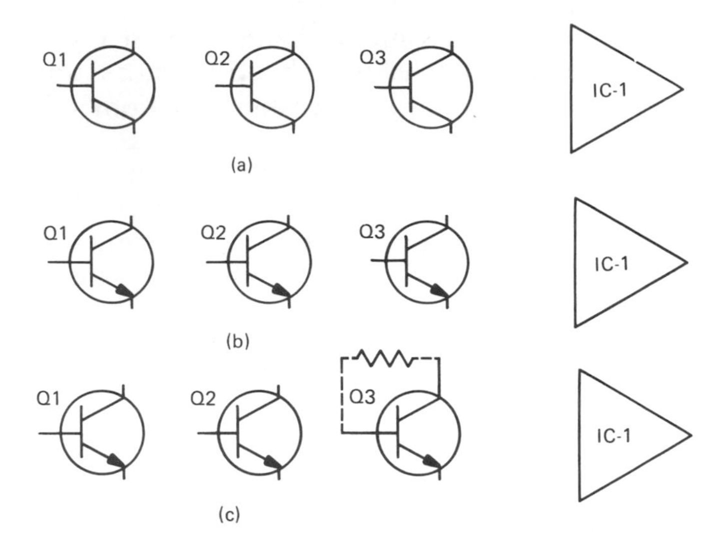

Note that the transistors are identified as NPN types on the basis of DC voltage and high-power ohmmeter measurements.

The integrated circuit, in this example, checked out normal on the basis of comparative DC voltage measurements.

On the other hand, Q3 showed abnormally high base-emitter bias voltage. A low-power ohmmeter check of Q3 showed comparatively low resistance from base to ground; this subnormal value raised the suspicion of leakage from collector to base.

An out-of-circuit check of Q3 with a low-power ohmmeter confirmed the suspicion of collector leakage, which measured 48,500 ohms.

Figure 5-1 Progressive troubleshooting "pictures." (a) Receiver employs three transistors and an integrated circuit; (b) transistors are identified as NPN types; (c) Q3 is found to have collector-base leakage.

crawl. On the other hand, the second set of resistance readings got around this difficulty, and the test data narrowed down the possible faults.

The possible faults were narrowed down by the in-circuit resistance readings with respect to the DC voltage readings—in other words, a high base-bias voltage on Q3, accompanied by a low base-to-ground resistance pointed to collector-base leakage.[1]

The trouble was then pinpointed when Q3 was checked

[1]This test data raises a strong suspicion of collector-base leakage, but in the absence of service data cannot be hard proof.

out-of-circuit with a low-power ohmmeter—its resistance from collector to base measured 48,500 ohms. When Q3 was replaced in the defective receiver, the sound output returned to normal, with full volume and without distortion. The DC voltages throughout the receiver were also restored to normal, on a comparative test basis.

Note that collector-base leakage does not make a resonance-probe check impractical, unless the leakage resistance is rather low. For example, a check at the collector of Q3 with a resonance probe (before the transistor was replaced), showed clearly that this was an IF circuit. The meter reading was 43 mV on open circuit (before the probe was applied), and dropped to 8 mV when the probe was applied. Next, when the generator was tuned to 455 kHz, the meter reading peaked up at 30 mV. After Q3 was replaced, the meter reding peaked up at 33 mV. However, when the collector-base leakage resistance is *less than 3 kilohms*, a resonance probe will *not* provide a clear peak indication.

Reminder

Don't forget that when there is suspicion that an open capacitor is causing a weak or dead trouble symptom, the suspected capacitor(s) can be quick-checked by temporarily bridging any suspect with a known good capacitor. Then, if normal reception is resumed, the troubleshooter knows that the capacitor under test is open and should be replaced.

COMPARATIVE DC VOLTAGES WITH AND WITHOUT SIGNAL

IF transistors in AM radio receivers are AVC biased, except that the last IF stage is sometimes operated with fixed bias. An AVC-biased transistor will normally show a bias-voltage shift from no-signal to strong-signal conditions. In turn, the troubleshooter can easily check for the possibility of AVC trouble when making comparative DC voltage checks. For example, the normal base and emitter DC voltages for an IF transistor in a small receiver are:

Without Signal	With Signal
Base, 1.050 V	Base, 0.935 V
Emitter, 0.368 V	Emitter, 0.030 V
Bias voltage = 0.682 V	Bias voltage = 0.905 V

Observe that the bias voltage on the NPN IF transistor in this example increases from 0.628 volt to 0.905 volt from the no-signal to the strong-signal condition of operation. This is an example of forward AGC bias. (See Figure 5-2.)

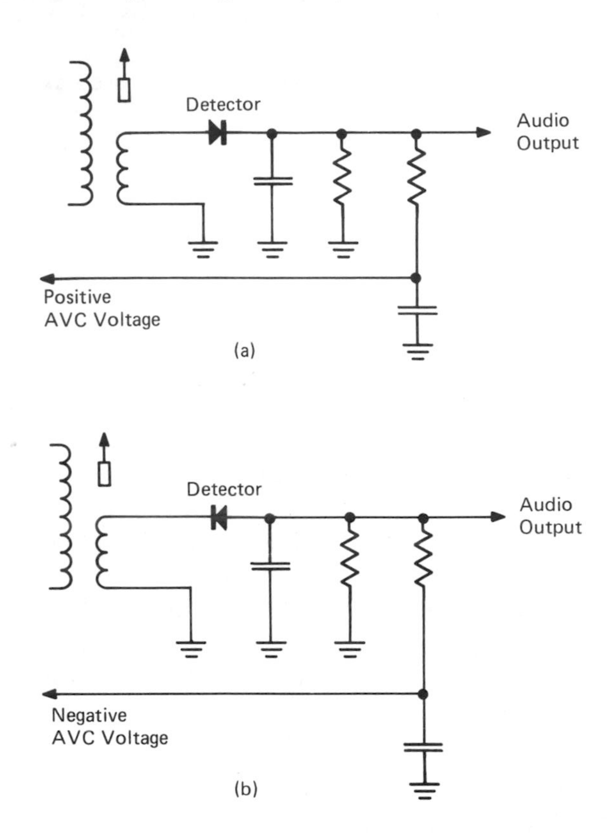

The configuration in (a) provides forward AVC bias for NPN IF transistors; it provides reverse AVC bias for PNP IF transistors. The configuration in (b) provides forward AVC bias for PNP IF transistors; it provides reverse AVC bias for NPN transistors.

Note: AVC action may be provided by forward-bias or by reverse-bias modes. When forward-AVC is used, the forward bias on an IF transistor normally increases with signal present. When reverse-AVC is used, the forward bias on an IF transistor normally decreases with signal present. Forward-AVC biases a transistor toward saturation; reverse-AVC biases a transistor toward cut off.

Figure 5-2 Basic AVC circuits. (a) Series detector, positive output; (b) series detector, negative output.

Thus, when making comparative DC voltage checks on similar AM radio receivers, the troubleshooter may need to check IF transistor bias voltages without signal, and also with signal present. In turn, if the AVC action in the bad receiver is widely different from the AVC action in the good receiver, the troubleshooter will conclude that the detector in the bad receiver is being driven by a subnormal IF signal, that the detector diode is defective, or that there is a component fault in the AVC circuit.

The IF output signal level can be checked with a rectifier probe and DVM; the audio output level can be checked on the AC function of a DVM; the resistance to ground in the AVC circuit can be measured with a low-power ohmmeter. Comparison measurements in the bad receiver and the good receiver guide the troubleshooter to the defective component or device.

Series vs. Shunt Detection

The troubleshooter will encounter shunt detectors occasionally. Basic series and shunt detector circuits are shown in Figure 5-3. A shunt detector may provide either positive-going output or negative-going output, in the same manner as a series detector. Transistor detectors will also be encountered, as depicted in Figure 5-4. The chief distinction between diode detection and transistor detection is that the latter provides a voltage gain, whereas the former imposes a voltage loss.

A transistor detector is essentially a combined diode detector and audio amplifier. It operates as a series detector (the base-emitter junction is connected in series with the secondary of the IF transformer). Either NPN or PNP transistor detectors may be utilized, depending upon whether positive-going AVC voltage, or negative-going AVC voltage is required.

Detector trouble symptoms are usually accompanied by AVC trouble symptoms, because the diode does double duty. However, an exception occurs in the case of separate AVC and detector diode configurations, as exemplified in Figure 5-5. When this arrangement is utilized, AVC trouble symptoms generally occur independently of detector trouble symptoms. Of course, if the IF transformer becomes defective, both the audio output and the AVC action will malfunction.

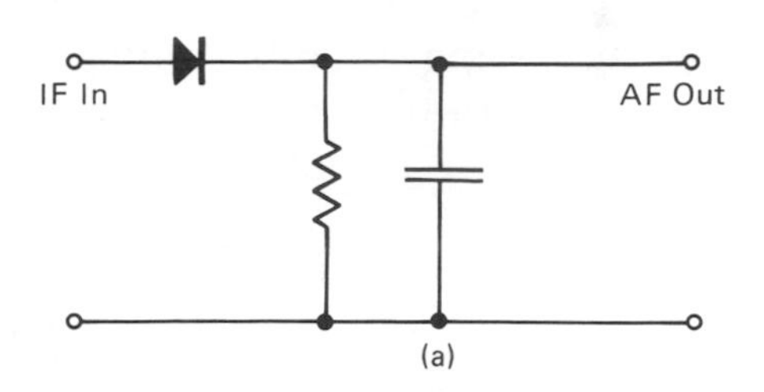

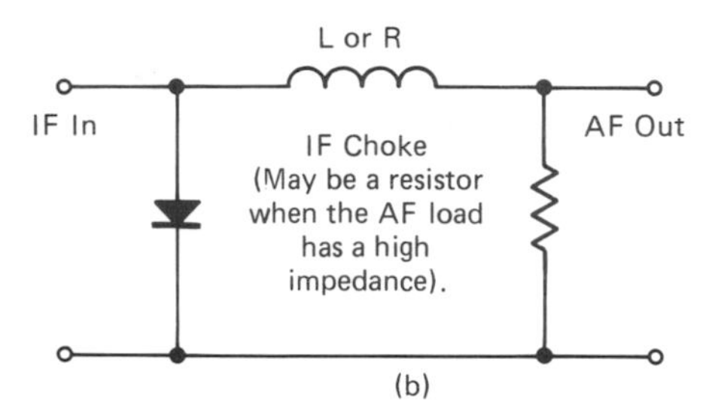

The series detector arrangement (a) is also called a voltage detector. It is the most widely used configuration in AM receivers. The shunt detector arrangement (b) is also called a current detector. It is used in some radio receiver designs, and in most demodulator probes.

Note: The shunt detector configuration has comparatively high output impedance. Either arrangement may employ either positive detection or negative detection. A peak-to-peak signal-tracing probe employs both series detection and shunt detection, with one diode operating as a positive detector, and the other diode operating as a negative detector.

Figure 5-3 Basic detector arrangements. (a) Series detector; (b) shunt detector.

Note on Circuit Disturbance

When making comparative DC voltage measurements, it is sometimes observed that connection of the DVM test leads into the receiver circuitry causes the trouble symptoms to change. This, of course, is an indication of objectionable circuit disturbance by the test setup. As an illustration, a voltage check in the converter circuit might cause the receiver system to buzz, to motorboat, or to squeal. Although this is not really new, it is very important:

The remedy for misleading test results due to circuit disturbance is to

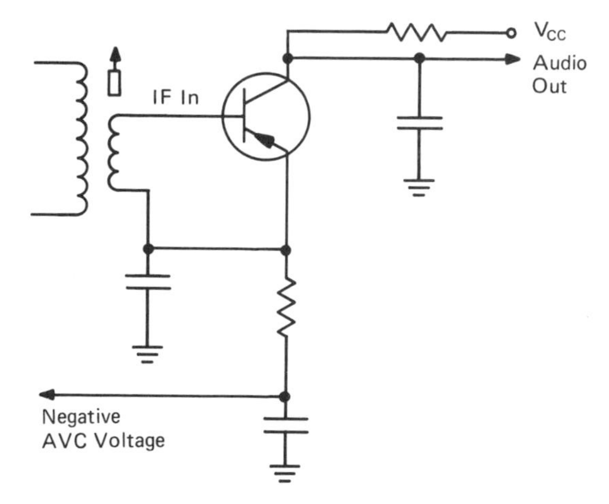

This arrangement provides forward-bias AVC for NPN IF transistors, or reverse-bias AVC for PNP IF transistors. If an NPN transistor detector is used, it will provide forward-bias AVC for PNP IF transistors, or reverse-bias AVC for NPN IF transistors.

Note: A transistor detector normally provides audio amplification in addition to IF detection. The transistor input circuit operates as a detector because it employs zero base-emitter bias. The transistor output circuit operates an an audio amplifier because the base is effectively driven by the demodulated IF signal and in turn the common-emitter configuration steps up this audio signal.

When replacing a defective IF transistor, keep in mind that a forward-AVC biased transistor has different characteristics, compared with a reverse-AVC biased transistor.

Figure 5-4 Basic transistor detector arrangement.

use a 100-kilohm isolating resistor in series with the "hot" test lead to the DVM. Since the input resistance of the DVM is 10 megohms, the resulting reduction in voltage readout is only 1 percent, and can be neglected.

When making voltage measurements in which both voltmeter leads are above ground in high-frequency circuitry, it is sometimes necessary to use two 100-kilohm isolating resistors—one resistor in series with the tip of each test lead. In this case, 200 kilohms are added in series with the measuring circuit; the resulting reduction in voltage readout is 2 percent, and this small experimental error can be neglected in practically all situations.

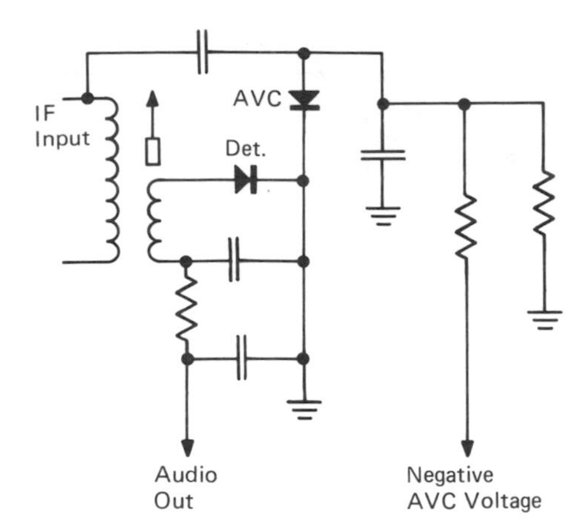

This configuration provides a negative AVC voltage. A positive AVC voltage is obtained by reversing the polarity of the AVC diode. Series detection is employed; the audio-output is pulsating DC and is negative-going. This pulsating DC will have its DC component removed, and will be changed into an AC waveform if a coupling capacitor is utilized in the audio-output lead.

Note: Radios designed for high-fidelity reproduction often employ separate AVC and detector diodes. Distortion is thereby minimized by reduction of AC shunt loading on the detector diode. The AVC diode is driven from the primary of the IF transformer, whereas the detector diode is driven from the secondary of the IF transformer.

Some radios designed for maximum sensitivity operate the detector diode with a slight forward bias. In turn, optimum demodulation is obtained in weak-signal reception.

Figure 5-5 Some receivers utilize separate AVC and detector diodes.

CONVERTER VOLTAGES IN OSCILLATORY AND NON-OSCILLATORY STATES

A practical example of DC voltage relations in converter circuitry is shown in Figure 5-6. The oscillator tuning capacitor was short-circuited in this example, and the receiver was dead, although a hissing sound could be heard from the speaker when the volume control was turned up to maximum.

Comparative DC voltage measurements at the converter transistor terminals in the bad receiver and the good receiver were as follows:

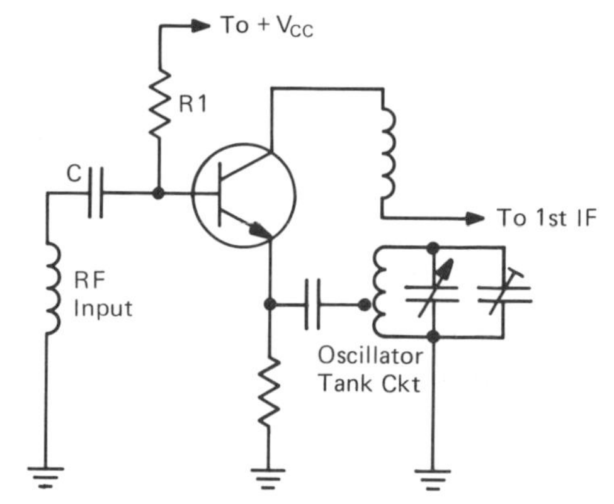

TROUBLE SYMPTOM: "DEAD RECEIVER"

Note: The base of the converter transistor is forward-biased by R1. However, this forward bias is masked by charge build-up on C when the local oscillator is operating. Oscillation normally occurs because the collector is inductively coupled to the oscillator tank, and positive feedback sustains oscillation.

This oscillation is normally sufficiently strong that the transistor conducts heavily on peaks of the oscillator waveform. Peak conduction results in spurts of electron flow in the base circuit; some of these electrons are stored on the right-hand side of C, and the average bias on the base becomes less positive.

In this particular example, oscillation caused the average bias on the base to be 0.54 volt (in the cutoff region).

When the oscillator tuning capacitor became short-circuited, oscillation was killed, and the bias on the base assumed a value of 0.66 volt (in the conduction region).

Figure 5-6 Typical oscillator circuitry in a converter stage.

Bad Receiver	Good Receiver
Collector, 4.81 V	Collector, 4.80 V
Base, 2.11 V	Base, 2.09 V
Emitter, 1.45 V	Emitter, 1.55 V

The emitter voltages show the greatest discrepancy, although the difference is not alarming.

However, the comparative voltages provide a definite trouble clue when bias values are noted. Thus, the base-emitter converter bias in the bad receiver is 0.66 volt, whereas the base-emitter converter bias in the good receiver is 0.54 volt. These bias voltages are evaluated as follows:

1. The bias voltage on the converter transistor in the bad receiver is in the conduction region.
2. The bias voltage on the converter transistor in the good receiver is in the cutoff region.
3. The converter transistor in the good receiver only *appears* to be cut off; it is not actually cut off because the local oscillator is operating, and the transistor conducts on the positive peaks of the oscillator waveform.
4. The converter transistor in the bad receiver is *not* cut off; *it is biased for class-A amplification.* Because the transistor is evidently workable, the logical conclusion is that there is a fault in the local-oscillator circuit which kills oscillation.
5. A follow-up RFI test between the two receivers confirms that the local oscillator in the bad receiver is either dead or far off frequency.
6. As noted above, the fault in the local-oscillator circuit was a short-circuited tuning capacitor.

All Voltages Negative

In the foregoing example, all of the voltages were positive, inasmuch as the negative side of the power supply was connected to ground (common). However, in the following example, all of the voltages are negative with respect to ground (common). The basic DC-voltage distributions that are involved in these examples are depicted in Figure 5-7.

In the following example, the bad receiver had the same fault as in the previous example—the local oscillator was killed, (due to a solder bridge that short-circuited the oscillator coil). Comparative DC voltage measurements at the converter transistor terminals in the bad receiver and in the good receiver were as follows:

Bad Receiver	Good Receiver
Collector, −7.96 V	Collector, −7.93 V
Base, −2.31 V	Base, −2.24 V
Emitter, −1.67 V	Emitter, −1.77 V

Note that the receiver in this example employed PNP transistors;

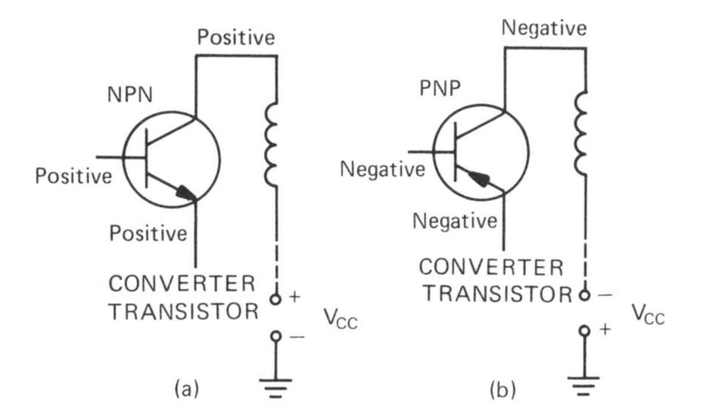

Note: All of the NPN transistor terminals are positive, because the negative side of the power supply is connected to ground. On the other hand, all the PNP transistor terminals are negative, because the positive side of the power supply is connected to ground.

Although this is the common practice, it is not the invariable rule, and the troubleshooter should keep the following exceptions in mind:

1. Occasionally, an NPN transistor will be operated with the positive side of the power supply returned to ground. In such a case, all of the NPN transistor terminals will be negative with respect to ground.
2. Occasionally, a PNP transistor will be operated with the negative side of the power supply returned to ground. In such a case, all of the PNP transistor terminals will be positive with respect to ground.

A converter transistor operates nonlinearly (a beat-frequency output cannot be produced if the transistor operates linearly). Therefore, the measured terminal voltages are affected, not only by the oscillator waveform, but also by the RF input signal waveform. This is just another way of saying that the bad receiver should be tuned to the same frequency as the good receiver.

Figure 5-7 Basic DC voltage distributions. (a) All voltages positive with respect to ground; (b) all voltages negative with respect to ground.

$+V_{CC}$ was returned to ground (common), and all of the transistor terminal voltages were negative with respect to ground.

The differences in DC voltages at the converter transistor terminals in the bad receiver and good receiver are not alarming. As in the case of the first receiver, however, the comparative voltages provide a definite trouble clue when bias values are noted. Observe that the base-emitter bias was 0.47 volt in the good receiver, whereas the base-emitter bias was 0.64 volt in the bad receiver.

The trouble clue is evident in the fact that the converter

transistor was biased into the cutoff region in the good receiver, whereas the converter transistor was biased into the conduction region in the bad receiver.

To recap the trouble analysis in this example, note that:

1. The converter transistor in the good receiver only *appears* to be cut off. It is cut off *on the average,* as "seen" by a DC voltmeter. In fact, the transistor conducts in spurts on the peaks of the oscillator waveform.
2. The pulse-type conduction of the converter transistor in the good receiver stores a charge on the base coupling capacitor, and on the emitter coupling capacitor. These stored charges oppose the fixed bias and shift the *average apparent bias* into the cutoff region.
3. There is no pulse-type conduction of the converter transistor in the bad receiver, and only the fixed bias voltage is measured. This fixed bias is 0.64 volt. In other words, the converter transistor in the bad receiver is biased for class-A amplification.[2]
4. This circuit-action analysis leads the troubleshooter to the conclusion that the converter transistor is probably in working condition, and that some circuit fault is killing the local-oscillator action.

OSCILLATOR SIGNAL SUBSTITUTION

Oscillator signal substitution is commonly provided by connecting the RF output from a signal generator to a small coil, and coupling this coil to the antenna loopstick in the bad receiver.

In turn, when the receiver is tuned to an AM station, and the generator is set to the station frequency plus 455 kHz, reception will be obtained in the event that the converter transistor is workable, but the local-oscillator function is dead. Thereby, preliminary conclusions can be confirmed or rejected.

---

[2]Note in passing that converter circuit action is obtained only when the converter transistor operates in class AB, class B, or class C; no beat output can occur in class A.

Similarly, the receiver may be tuned to an FM station, and the output from an RF generator coupled into the FM mixer circuit. Note that the difference frequency is 10.7 MHz in this situation. This oscillator signal-substitution test should be made only with a lab-type VHF signal generator, however.

MISCELLANEOUS QUICK-CHECK EXAMPLES

Comparative current-drain tests are sometimes informative.

For example, a current meter can be connected in series with the V_{CC} battery in the bad receiver and in a similar good receiver. A typical good AM/FM pocket radio has a current demand of approximately 10 mA at low volume output, and about 25 mA at high volume output. If the bad receiver happens to draw 10 mA at low volume, but draws only 15 mA when the volume control is advanced to maximum (and has weak output), the troubleshooter would look for a malfunction in the audio section.

In another example, a typical good AM/FM portable radio has a current demand of approximately 11 mA at low volume output, and up to 30 mA at high volume output. If the bad receiver happens to draw 5 mA at low volume, and draws only 14 mA when the volume control is advanced to maximum (and has weak and/or distorted output), the troubleshooter would look for a malfunction in the RF or IF section.

Functional comparative current-drain tests are occasionally useful.

For example, an FM/AM portable radio normally has a current demand of 10 mA on its AM function at low volume, and draws 11 mA on its FM function at low volume. Then, if the bad receiver draws 10 mA on its AM function, and 20 mA on its FM function at low volume, the troubleshooter will conclude that the trouble will be found in a stage that is operative only on the FM function.

Relative oscillator-injection voltage measurements can finger a poor-sensitivity trouble symptom in some situations.

As an illustration, normal converter gain depends upon the local-oscillator injection voltage (Figure 5-6). In other words, the converter transistor is both an oscillator and a heterodyne mixer. The incoming RF signal is mixed with the local-oscillator voltage in the converter transistor, and the resulting 455-kHz beat frequency is developed at the collector.

If the oscillator-injection voltage is weak, the result is a poor-sensitivity trouble symptom. (The cause of weak oscillator-injection voltage can be a leaky or open capacitor, for example.)

A relative oscillator-injection voltage measurement is a comparison quick-check made with respect to a similar good receiver. The check is made at the collector of the converter transistor, using a peak-reading probe and a DVM. For example, 0.6 volt is a typical ball-park value. A substantially lower value would indicate malfunction in the circuitry.

Experiment

This experiment continues with the radio kit project described in Chapter 4. Follow the steps listed in the construction manual, and align the tuned circuits according to instructions. Then, make the following measurements and note the values on the schematic diagram:

1. DC voltages at the terminals of TR1, TR2, and IC LM-386.
2. Resistances to ground from the terminals of TR1, TR2, and IC LM-386, using a low-power ohmmeter.
3. Signal voltage at the input to D1, using a peak-reading probe and DVM. (Receiver should be driven from an AM signal generator, although the experiment can be made with the receiver tuned to a station signal.)
4. AVC voltage across C13, with receiver tuned to a strong station signal, and then tuned to a weak station signal.

6

Progressive Radio Troubleshooting Techniques

*Transceiver Quick Checks * Comparative Field-Strength Test * Audio Section Malfunction * Measurement of Quench Frequency * Measurement of High-Frequency Voltage at Antenna * Incidental Frequency Modulation * CB Walkie-Talkie Quick Checks * CB Walkie-Talkie RFI * Check of Modulation Waveform * Test Tips * Experiment*

TRANSCEIVER QUICK CHECKS

Radio transceiver troubleshooting is generally facilitated by the circumstance that transceivers are used in similar pairs. Usually, only one unit becomes defective at a time, and it can be checked against the good unit in comparison tests. (See Chart 6-1.)

With reference to Figure 6-1, the smallest transceivers employ superregenerative detectors for reception. This type of transceiver normally operates at approximately 50 MHz, or more precisely, at 49.860 MHz. When the trouble symptom is weak or no reception, comparison DC voltage measurements should start at the base, emitter, and collector terminals of the detector transistor. As an illustration, the following voltages are typical: base, 0.503 V; emitter, 0.040 V; collector, 2.01 V.[1]

As shown in Figure 6-1, an NPN transistor is employed in this example; the average base-emitter bias voltage is 0.463 volt. The bias voltage is developed by RF and quench oscillator action. Note that if the transistor does not oscillate, its base-emitter bias will be zero. If the transistor terminal voltages are substantially off-value in a

[1]It is good practice to use a 100 kilohm isolating resistor with the DVM.

CHART 6-1

Typical Low-Power Transceiver Functional Arrangement

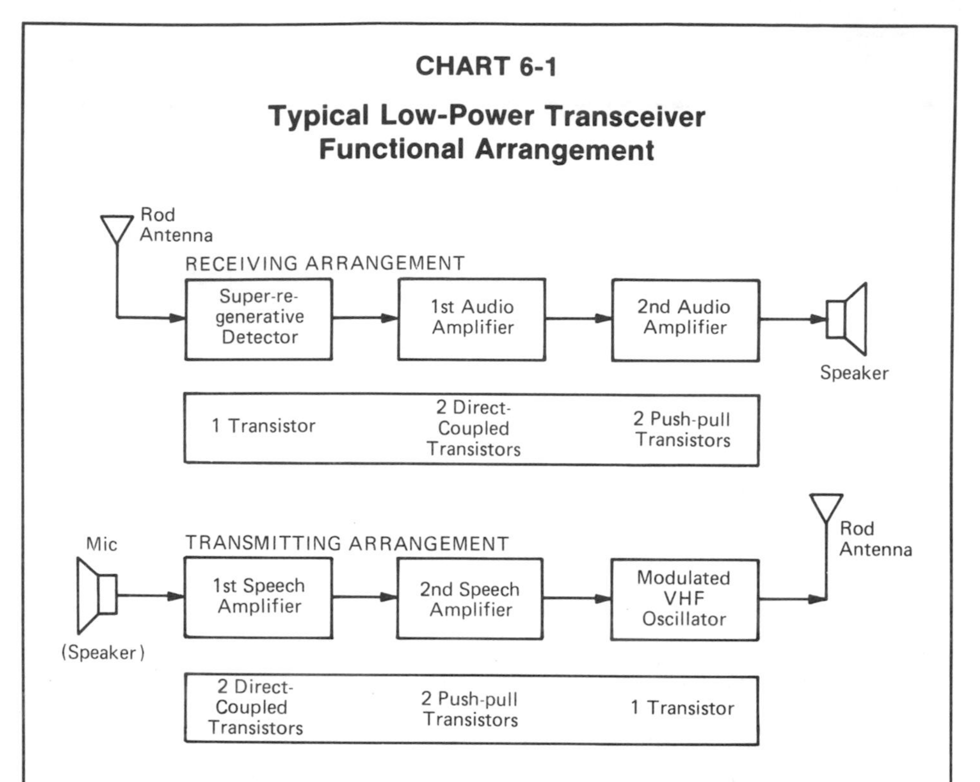

Five transistors are employed in this representative low-power transceiver arrangement. It operates at a fixed frequency of 49. 890 MHz. The same transistors are utilized for reception and for transmission. A push-to-talk switch rearranges the circuitry as shown in the block diagrams for changeover from reception to transmission. Note that the speaker serves as a microphone during transmission. The VHF oscillator is crystal-controlled.

When troubleshooting without service data, the first and second audio stages can be easily identified by means of scope checks in the good transceiver. Transistor types and basing can be determined by means of DC voltage measurements. Remember that no detector diode is used in this type of receiver—a diode is often present, but it functions as a bias stabilizer for the push-pull transistors.

Do not attempt to operate the transceiver in its transmitting mode unless the rod antenna is fully extended. Otherwise, the transistor in the modulated VHF oscillator stage will be forced to dissipate excessive power in the form of heat—and the transistor is apt to burn out.

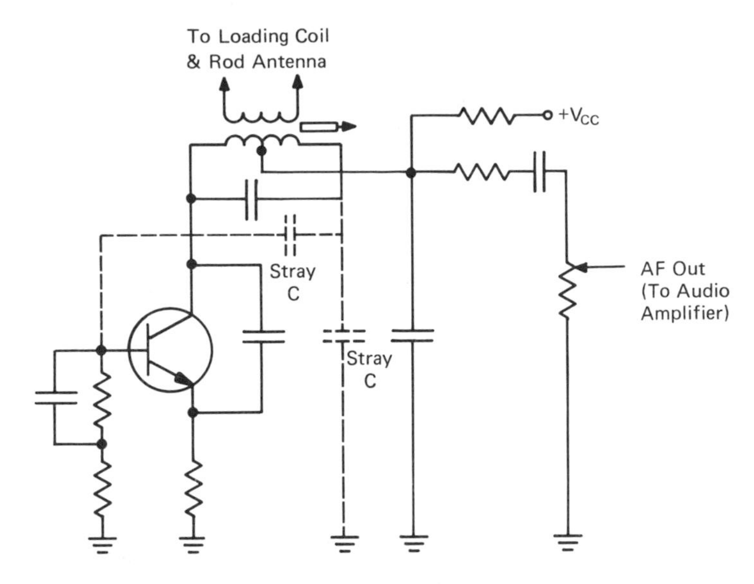

Regardless of the trouble symptom, it is advisable to start by measuring the resistance from V_{CC} to ground with a low-power ohmmeter. (Approximately 20 kilohms is normal, in this example.) This test will catch many leaky capacitors that can cause tough-dog trouble symptoms.

Note: A superregenerative detector is an oscillator in which the amplitude of oscillation is controlled by the incoming signal voltage. The transistor oscillates at the signal frequency and also at a supersonic frequency called the quench frequency. The quench frequency is produced by blocking-oscillator action. A large amount of positive feedback is employed, and the RF oscillation voltage builds up rapidly, accompanied by a rapid increase in collector current. However, before the RF oscillation voltage can build up to the saturation point, the quench action drives the transistor momentarily into cutoff. Then, the RF oscillation voltage build-up starts again. Since the starting point of the build-up is from the peak of the instantaneous signal voltage, the oscillation build-up attains a higher amplitude when the signal voltage has a higher amplitude. The chief advantage of a superregenerative detector is that it provides very high sensitivity with only one transistor.

Figure 6-1 Typical 50-MHz superregenerative detector configuration.

comparison check, additional test data can be obtained by measuring circuit resistances to ground with a low-power ohmmeter.

On the other hand, if normal voltages are measured in the detector stage, a trouble symptom of weak or no reception is most likely to be tracked down to an audio-amplifier malfunction.

Note that transceivers such as that exemplified in Figure 6-1 generally use the same transistor for both superregenerative detection in reception, and for RF output in transmission. During transmission, the transistor is switched for operation as a modulated quartz-crystal oscillator. Accordingly, useful test data can often be obtained by cross-checking the transistor terminal voltages with the transceiver in its transmitting mode. For example, when the arrangement in Figure 6-1 is switched into its transmitting mode, the following voltages are typical: base, 0.308 V; emitter, 0.484 V; collector, 7.69 V.

> ***Caution:*** **The transceiver should not be operated in its transmitting mode unless the rod antenna is fully extended—the transistor is likely to burn out if it is required to dissipate the normally radiated RF energy in the form of heat.**

As a practical note, this type of transceiver normally outputs a comparatively loud hissing sound during reception, if there is no input signal. (It is analogous to an FM receiver in this respect.) When an input signal is present, the detector is quieted and the hissing sound ceases.

COMPARATIVE FIELD-STRENGTH CHECK

The radiated field strength of a transceiver can be compared with that of another transceiver by means of a simple peak probe and DVM test. For example, if a test lead is suspended as an antenna and one end is connected to a peak-reading probe and DVM, a meter reading is normally obtained. (See Figure 6-2.)

When the test lead is spaced by one yard from the rod antenna of an Archer (Radio Shack) walkie-talkie, three typical meter readings are as follows:

1. With the walkie-talkie turned off, the DVM indicated 8 mV; this reading resulted from pickup of stray electromagnetic fields by the test lead.
2. With the walkie-talkie turned on and set for reception, the DVM indicated 9 mV; the additional 1-mV reading resulted from re-radiation of the superregenerative detector.
3. With the walkie-talkie turned on and set for transmission, the DVM indicated 50 mV; the additional 42-mV reading resulted from carrier radiation by the transmitter.

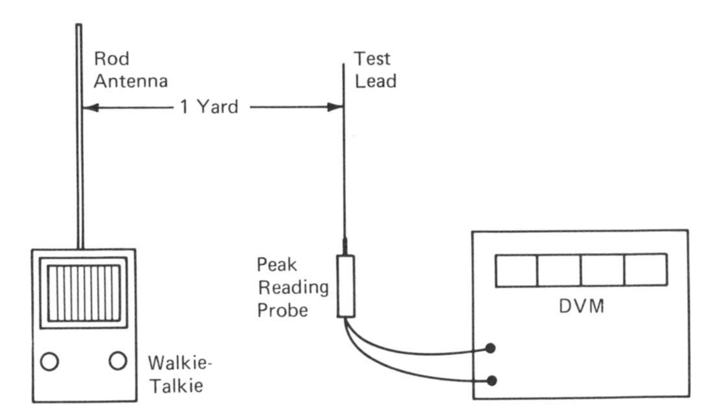

This arrangement is very useful for comparative field-strength quick checks of low-power walkie-talkies. The test lead may be suspended on the wall back of the bench. A small field-strength reading is usually noted when the walkie-talkie is turned off, due to stray field pickup from broadcast stations, television stations, amateur radio stations, CB transmissions, and so on. This background field-strength indication will vary considerably from one location to another.

Note that an unlicensed electronic troubleshooter may legally make adjustments to the transmitter in a 49.860-MHz walkie-talkie. On the other hand, adjustments to the transmitter in a 27-MHz walkie-talkie may legally be made only by, or under the direct supervision of, a licensed first- or second-class radio operator.

Figure 6-2 Relative field-strength test arrangement.

Note that this transceiver is rated for 100 mW RF power in its transmitting mode. This means that the power input to the oscillator transistor is normally 100 mW (less RF power will actually be radiated). Thus, with a collector DC voltage of 7.69 volts, the collector DC current flow will normally be approximately 13 mA. Note also that this transceiver is rated for about 25 mA battery current demand in the receiving mode, and about 35 mA battery current demand in the transmitting mode. Thereby, two additional and useful quick checks are available.

Audio Section Malfunction

As a practical troubleshooting angle, note that the audio section in this type of transceiver employs the audio transistors as a *speech amplifier* during transmission. This means that if there is a malfunction in the audio section, it will be evident during both reception and transmission. (If the audio section is dead, there will be no reception—carrier output will be obtained during the transmission mode, but no modulation will be provided.)

MEASUREMENT OF QUENCH FREQUENCY

It is sometimes helpful to check the quench frequency in a superregenerative detector. Receiver sensitivity becomes poorer if the quench frequency is too low or too high with respect to the RF signal frequency. In the case of the 50-MHz receiver exemplified above, the normal quench frequency is approximately 300 kHz.

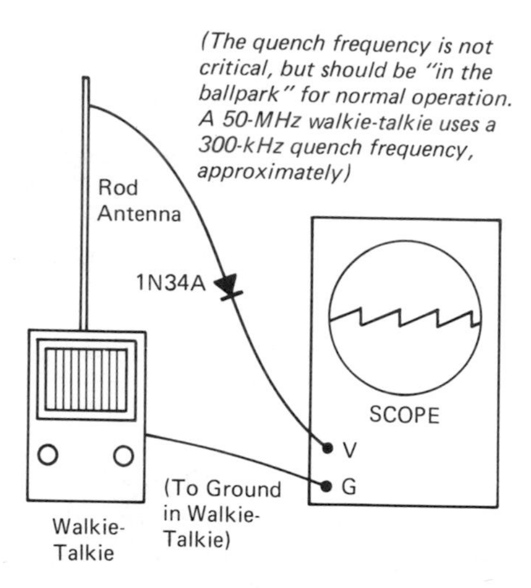

The re-radiated RF voltage from the superregenerative detector is demodulated by the 1N34A diode and normally appears as a sawtooth waveform on the scope screen. This is the modulation envelope of the quench activity. The quench frequency is measured by means of the calibrated time base in the scope, or the pattern frequency can be checked with an AM signal generator.

To check the pattern frequency with an AM signal generator, substitute the generator output signal for the walkie-talkie signal input to the scope. Then, adjust the generator frequency to match the number of cycles that were displayed in the sawtooth pattern. In turn, the generator indicates the quench frequency.

Figure 6-3 Check of quench frequency.

To measure the quench frequency, connect the rod antenna through a 1N34A diode to the input of an oscilloscope, as shown in Figure 6-3. The scope may be operated on its DC function to minimize hum interference; the pattern will be displaced vertically by the rectified DC output from the diode, but the pattern can be recentered on the screen by adjustment of the vertical positioning control.

The quench voltage will normally be displayed as a sawtooth pattern on the scope screen. Its frequency can easily be measured if the scope has a calibrated time base. An off-frequency quench voltage points to a failing component in the superregenerative detector circuit. Leaky capacitors are ready suspects.

MEASUREMENT OF HIGH-FREQUENCY VOLTAGE AT ANTENNA

Another informative comparison test for low-power walkie-talkies is shown in Figure 6-4. A peak-reading probe is applied near the base of the rod antenna, and the probe output is fed to a DVM. Thereby, the high-frequency voltage at the antenna of a good transceiver can be compared with the high-frequency voltage at the antenna of a bad receiver.

Typical readings for a good 100-mW 50-MHz transceiver are as follows:

Transceiver Turned Off	Transceiver Turned On (Reception)	Transceiver Turned On (Transmission)
13 mV	71 mV	141 mV

Note that a reading was obtained when the transceiver was turned off, due to stray pickup of fields from broadcast stations, TV stations, CB transmissions, amateur radio stations, and so on. The DVM reading increased by 58 mV when the transceiver was turned on for reception, due to the re-radiation from the superregenerative detector. Then, when the transceiver was switched to its transmitting mode, the radiated carrier voltage increased the DVM reading by 70 mV.

Note that when two superregenerative receivers are placed close together, their noise output increases considerably at a certain spacing. This increase in noise results from the fact that each of the superregenerative receivers re-radiates its own noise

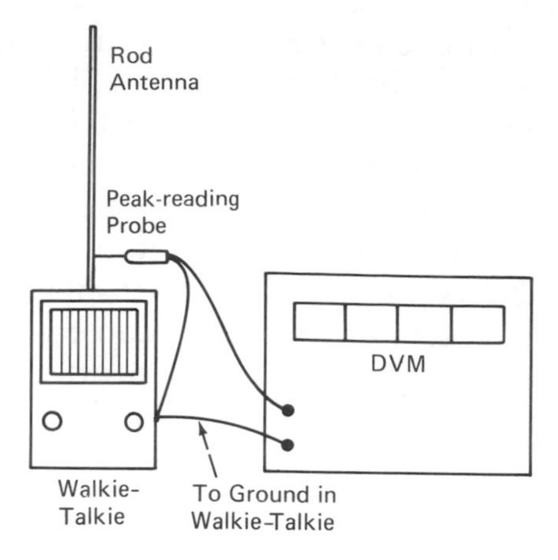

Note: Similar walkie-talkies in normal operating condition will not provide precisely the same high-frequency voltage output, although the DVM readings will not vary more than 10 percent, approximately. On the other hand, a walkie-talkie that exhibits a weak-transmission trouble symptom might show only one-quarter of the high-frequency voltage provided by a good walkie-talkie. In such a case, the troubleshooter concludes that the fault will be found in the high-frequency circuitry.

On the other hand, a walkie-talkie that exhibits a weak-transmission trouble symptom might show as much high-frequency voltage as that provided by a similar good walkie-talkie. In this case, the troubleshooter concludes that the trouble will be found in the modulator circuitry.

Figure 6-4 Check of high-frequency voltage at antenna.

signal which in turn is picked up by the other receiver and added to its normal noise level.[2]

Note also that when you are working with two transceivers located near each other, with one transceiver in its receiving mode and the other transceiver in its transmitting mode, acoustic feedback will take place and produce a loud howl. This is normal response, and does not represent a trouble symptom.

INCIDENTAL FREQUENCY MODULATION

You will observe that a walkie-talkie can cause radio-frequency interference (RFI) to an FM receiver nearby (see Figure 6-5).

[2]This effect occurs only when the spacing between the receivers is suitable for a required level of re-radiation pickup.

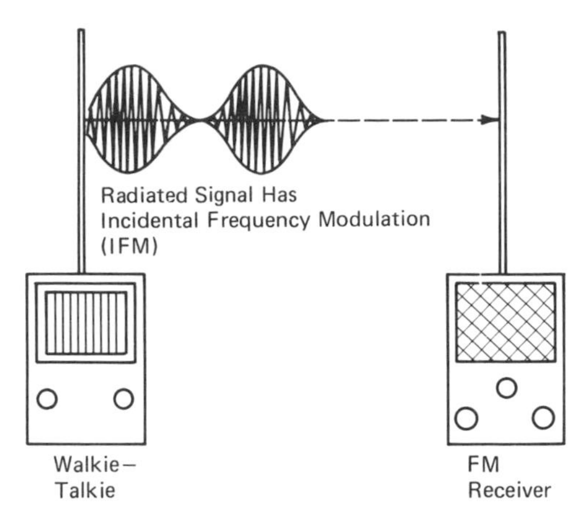

An FM receiver does not respond to amplitude modulation. It responds only to frequency modulation. Note that the radiated signal from a walkie-talkie is reproduced by an FM receiver because of incidental frequency modulation. An amplitude-modulated oscillator produces an AM output and an IFM output. An FM receiver responds to the IFM component in the radiated signal.

Note: An amplitude-modulated oscillator produces incidental frequency modulation as a result of junction capacitance. In other words, as the collector voltage of the oscillator transistor swings up and down, the collector-base junction capacitance varies accordingly. This capacitance variation swings the frequency of the oscillatory circuit higher and lower in accordance with the collector-voltage variation. The result is more or less incidental frequency modulation accompanying the amplitude modulation.

Note that if the walkie-talkie operates at 49.860 MHz, the FM receiver must be tuned to 99.72 MHz to receive the transmission from the walkie-talkie. Stated otherwise, second-harmonic reception is used by the FM receiver.

Figure 6-5 Walkie-talkie transmission can be heard on an FM receiver.

This interference normally has a comparatively low level—it is not a trouble symptom unless the interference is received at an appreciable distance. An abnormally high RFI level can be caused by improper operation of the modulated oscillator in the walkie-talkie.

In ideal operation, the modulated oscillator would generate a true sine carrier wave, and the carrier would still have a true sine waveshape at maximum percentage of modulation. In practice, the output from the modulated oscillator is less than ideal—with reference to Figure 6-6, the carrier generally has more or less second-

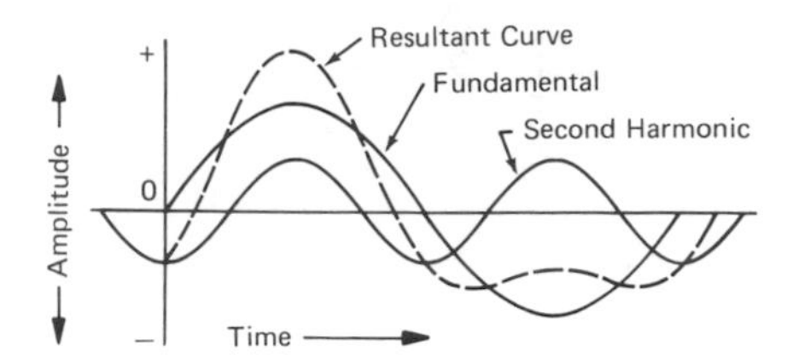

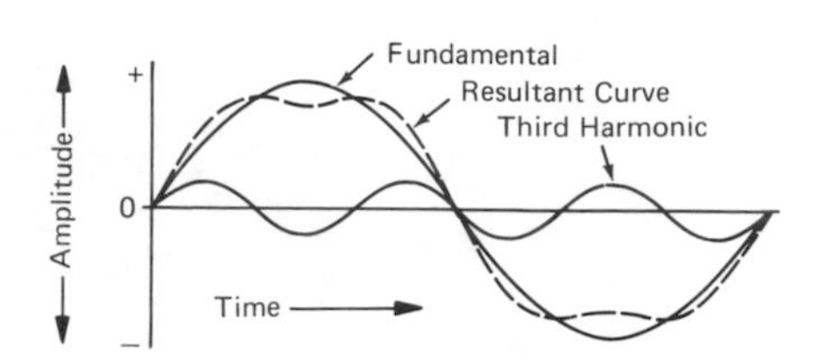

Observe that second-harmonic distortion of a waveform produces an unsymmetrical resultant—its peak voltages are unequal. The resultant waveform is also nonsinusoidal. On the other hand, third-harmonic distortion produces a symmetrical resultant—its peak voltages are equal. However, this resultant is also nonsinusoidal.

In other words, either even-harmonic or odd-harmonic distortion of the carrier causes its waveshape to become nonsinusoidal.

Normally, the percentages of second-harmonic and of third-harmonic components in the radiated waveform from a walkie-talkie are comparatively small. However, abnormal operating conditions in either the modulated-oscillator section or in the modulator section can greatly increase the percentages of the second-harmonic and third-harmonic components.

Carrier distortion in a modulated-oscillator stage seldom produces second-harmonic distortion only, or third-harmonic distortion only. In practically all cases of malfunction, both second-harmonic and third-harmonic distortion are abnormally increased. Note also that higher-order even and odd harmonic components are usually generated, such as fourth and fifth harmonics. However, higher-order harmonic voltages are ordinarily comparatively weak.

Figure 6-6 Basic forms of harmonic distortion.

harmonic and third-harmonic distortion. The percentage distortion increases at 100-percent modulation in most cases, and increases greatly in the case of overmodulation.

Therefore, when a comparative check shows that a walkie-talkie is radiating excessive RFI, the troubleshooter turns his attention to the modulated oscillator. DC voltages and resistances to ground can be measured to pinpoint most oscillator-circuit faults (capacitors are the chief exception). The audio-frequency modulator waveform can be checked in the good and bad transceivers by means of an oscilloscope. For example, if a modulator overdrives a modulated oscillator, splatter is produced, with a resulting large increase in RFI.

CB WALKIE-TALKIE QUICK CHECKS

Walkie-talkies with power ratings of more than 100 mW operate in the CB from 26.965 to 27.405 MHz. Most of the quick checks that have been explained for 50-MHz walkie-talkies are also applicable to 27-MHz walkie-talkies. One basic exception is in the fact that superheterodyne circuitry is used instead of superregenerative circuitry; an RF amplifier stage is also typically included in a CB walkie-talkie. (See Chart 6-2.)

In most cases, the first check of a bad CB walkie-talkie should be a DC current measurement at the battery clip (or power supply terminals). As an illustration, a 300-mW CB walkie-talkie is rated for a receiving no-signal current drain of 26 to 37 mA, and for a transmitting no-modulation current drain of 100 to 150 mA.

Substantially larger or smaller current drains would indicate resistive faults in the V_{CC} branch circuitry. If the current drains are reasonable, a follow-up check of the V_{CC} system impedance should be made. An audio impediance checker is used. For example, when the Z checker is connected across the battery-clip terminals in a good 300-mW CB walkie-talkie, a DVM reading of 14 millivolts is typically obtained when the audio oscillator is set for 5 volts output at 60 Hz.

A check of the V_{CC} impedance is advisable at the start of troubleshooting procedures, because open capacitors or high power-factor capacitors in the V_{CC} network can cause various tough-dog trouble symptoms.

CB WALKIE-TALKIE RFI

The 27-MHz oscillator is not amplitude-modulated in a typical 300-mW CB walkie-talkie; amplitude modulation takes place in the final amplifier following the crystal oscillator. Nevertheless, there is normally a small amount of incidental frequency modulation. There is also some harmonic output. In turn, when the normally operating walkie-talkie is located near an FM receiver, transmission is picked up at 108.5 MHz on the tuner (the fourth harmonic of the walkie-talkie Channel 14 transmitting frequency).

Excessive RFI points to a fault in the final-amplifier or modulator circuitry. As noted previously, one of the most common

CHART 6-2

Typical 300-mW CB Walkie-Talkie Functional Arrangement

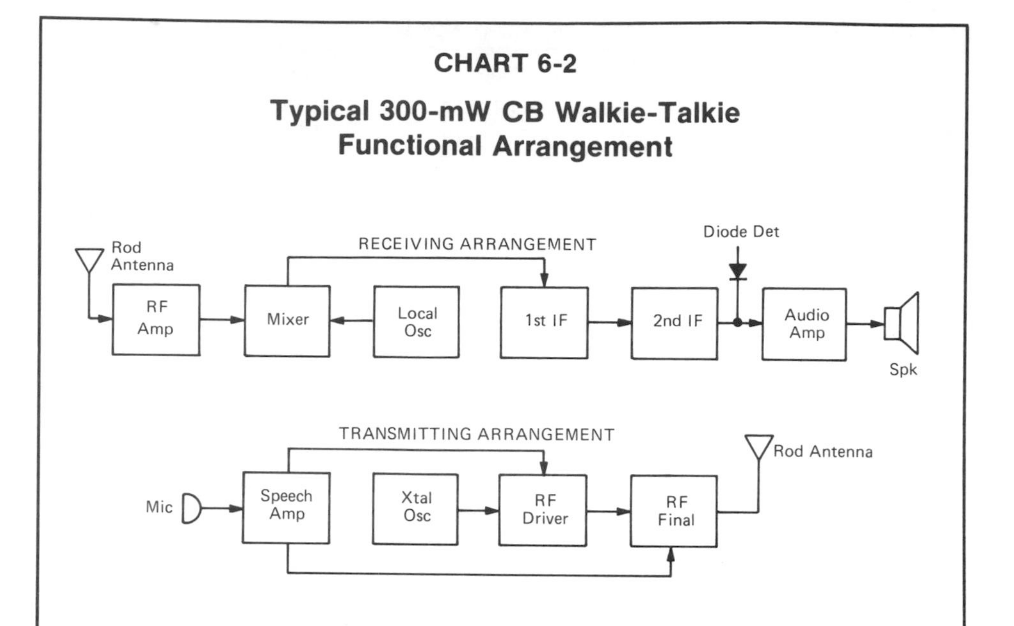

The receiver employs five transistors, a diode detector, and an IC audio amplifier.

The transmitter utilizes three transistors and an IC speech amplifier.

In this example, the same IC is used for audio amplification and for speech amplification. The transmitter transistors are separate from the receiver transistors.

Note that the microphone and the speaker are separate units in this type of walkie-talkie. The speech amplifier is also the modulator section; it is transformer-coupled to the RF driver and the RF final—both the driver transistor and the final transistor are amplitude-modulated by variation of collector voltage.

Separate quartz crystals are used in the local-oscillator section and in the transmitting XTAL-oscillator section. The intermediate frequency is 455 kHz.

When troubleshooting without service data, the audio section is easily identified as an integrated circuit. Transistor types and basing can be determined by means of DC voltage measurements. Since walkie-talkies are generally used in pairs, a good unit is almost always available for comparison tests with the bad unit. This situation greatly facilitates troubleshooting procedures.

Do not attempt to operate the walkie-talkie in its transmitting

CHART 6-2 CONTINUED

mode unless the rod antenna is fully extended. Otherwise, the transistor in the RF final stage will be forced to dissipate excessive power in the form of heat—and the transistor is apt to burn out.

causes of excessive RFI from walkie-talkies is overmodulation and resulting "sideband splatter." Oscilloscope checks in the bad and good walkie-talkies can provide highly informative test data in the case of RFI trouble symptoms.

CHECK OF MODULATION WAVEFORM

A good transceiver can be used with a service oscilloscope to check the modulation waveform from a bad transceiver, as shown in Figure 6-7. The bad transceiver should be tone-modulated by the sound output from a speaker driven from an audio oscillator, for example. If a high-fidelity speaker is used, the sound output will have a reasonably good sine waveform, provided that the audio oscillator supplies a good sine waveform.

A 1-kHz audio tone is suitable. Its level should be adjusted to obtain normal loudness of output from the good transceiver placed two or three yards distant. The modulation waveform depicted in Figure 6-7 has a true sine envelope, and is an example of ideal amplitude modulation. On the other hand, a bad transceiver may radiate a distorted sine envelope. Or, the amplitude of modulation may exceed 100 percent, with resulting clipping of the "trough" interval in the modulating waveform.

A bad transceiver may be incapable of 100-percent modulation. In some cases, the upper half of the modulation envelope may have a different waveshape with respect to the lower half of the modulation envelope. Again, the peaks of the upper and/or lower modulation envelope may be compressed or clipped.

In any case, a knowledge of the modulation waveform that is being radiated by the bad transceiver can be of considerable assistance to the troubleshooter. The displayed distortion(s) provide clues to the location of the fault; moreover, the scope serves as a highly informative monitor, to show whether or not a particular adjustment or replacement has improved the modulation waveform.

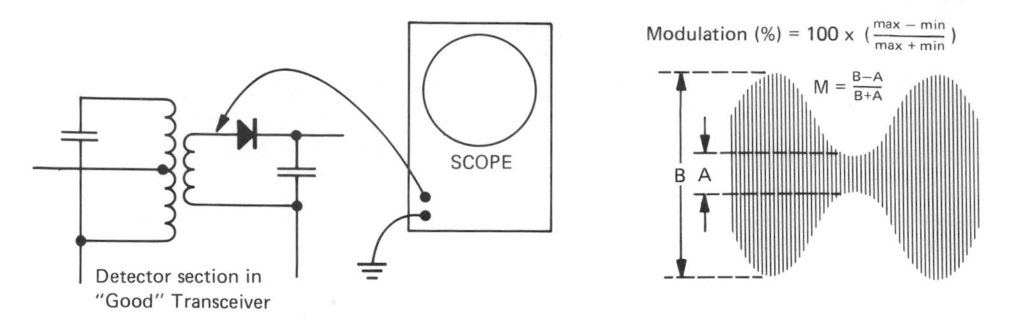

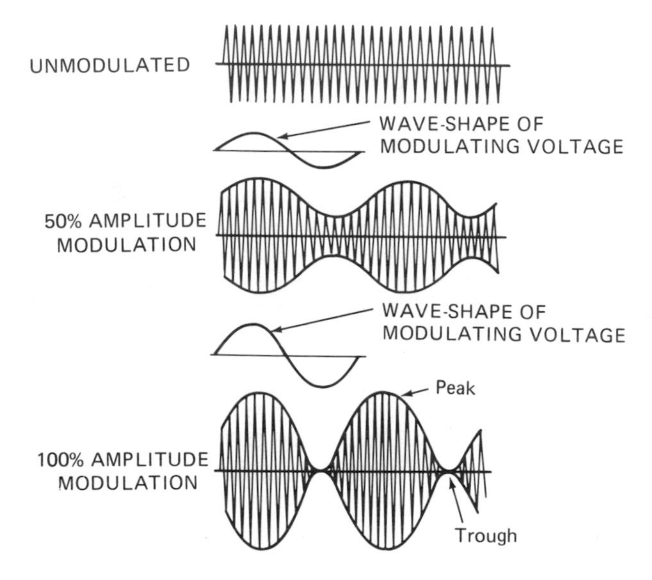

The signal radiated from the antenna of the bad transceiver is reproduced by the good transceiver.

To check the modulation waveform produced by the bad transceiver, a scope is connected at the input of the detector in the good transceiver. In turn, the modulation waveform is displayed on the scope screen.

Note that the modulated signal radiated by the bad transceiver has a carrier frequency of 27 MHz (which cannot be displayed by most scopes). On the other hand, the modulated signal reproduced at the input to the detector in the good transceiver has a carrier frequency of 455 kHz, which can be displayed by any service scope.

Figure 6-7 Good transceiver used to check modulation waveform of bad transceiver.

TEST TIP REMINDERS

Although these test tips are not really new, they are of basic importance in practical troubleshooting procedures, and are recapped as reminders:

Distorted Modulation: Look for a malfunction in the modulator section. The microphone could be defective—try another known good microphone.

Transmitter Operates with XTAL Unplugged: The oscillator, driver, or final circuits are self-oscillatory. The most likely culprit is an open decoupling or bypass capacitor in these sections.

Transmitter is Off-Frequency: This malfunction is almost certain to be caused by a defective quartz crystal.[3]

Weak Transmitter: RF amplifier transistors are ready suspects, followed by defective capacitors. (The final transistor can be damaged if the rod antenna is not fully extended.)

Intermittent Reception: Intermittents may be thermal, mechanical, or triggered by transient voltages. Check switches, defective insulation, solder joints, pressure contacts, and variable controls. Transistors, diodes, resistors, and crystals may become internally intermittent.

Overload on Strong Signals: When the receiver overloads and distorts on strong-signal reception, the AVC circuit is most likely to be defective. Leaky capacitors are the most common culprits.

Weak Reception: When reception is weak, the trouble could be as simple as a poor antenna connection. Otherwise, look for a low-gain stage or a defective detector diode. Signal-injection tests made on a good transceiver for comparison with responses of the bad transceiver are invaluable.

Experiment

This experiment demonstrates the responses of basic types of radio receivers to a small noise generator, like that used in preliminary troubleshooting procedures:

---

[3] A quartz crystal occasionally becomes intermittent and may not oscillate unless the power switch is turned on and off several times. Occasionally, a quartz crystal will suddenly "jump" to another frequency—the crystal must be replaced.

1. Turn on an AM radio receiver and adjust the tuner between channels. Hold the noise generator near the receiver and turn on the generator. Note that a *loud* rushing sound is heard from the speaker when the receiver's volume control is advanced.
2. Turn on an FM radio receiver and adjust the tuner between channels. Hold the noise generator near the rod antenna and turn on the generator. Note that the speaker remains *completely silent.* (An FM receiver normally rejects amplitude-modulated signals.)
3. Turn on a 27-MHz CB radio receiver. Hold the noise generator near the rod antenna and turn on the generator. Note that a *moderately loud* rushing sound is heard from the speaker when the receiver's volume control is advanced.
4. Turn on a 50-MHz CB radio receiver. Hold the noise generator near the rod antenna and turn on the generator. Note that a rushing sound is heard from the speaker when the receiver's volume control is advanced, but that the sound is *not as loud* as that from the 27-MHz receiver.

The sound output was greater in the test of the AM radio than in the test of the 27-MHz CB receiver because the CB receiver responded to higher harmonics in the noise signal. The sound output in the test of the 27-MHz CB receiver was greater than in the test of the 50-MHz CB receiver because the 50-MHz receiver responded to still higher harmonics in the noise signal. (The higher harmonics in a noise-generator signal become progressively weaker.)

AM/FM Noise Injector

Next, construct the AM/FM noise injector shown in Figure 6-8. The parts can be assembled in a miniature experimenter box. Repeat the foregoing noise-signal tests, and note that this type of noise generator will energize an FM receiver.

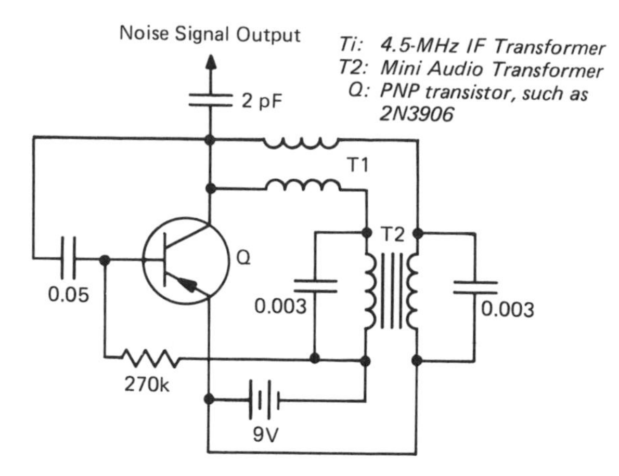

Note: This noise injector has both AM and FM output because it oscillates at both audio frequency and at 4.5 MHz. The 4.5-MHz signal is both amplitude-modulated and frequency-modulated by the audio signal voltage. In other words, as the collector voltage rises and falls with the audio signal voltage, the collector-base junction capacitance increases and decreases accordingly. In turn, the resonant frequency of T1 rises and falls correspondingly. As a result, the signal output is both amplitude-modulated and frequency modulated. The primary and secondary windings on T1 and T2 are tightly coupled, with the result that a large amount of feedback is provided; in turn, the signal waveform is nonsinusoidal and has a comparatively high harmonic content.

Hint: If no RF oscillation is obtained, reverse the connections to the secondary of T1. If no AF oscillation is obtained, reverse the connections to the secondary of T2.

Figure 6-8 An AM/FM noise-injector configuration.

7

Television Troubleshooting Without Service Data

*Preliminary Approach * Functional Sections Associated with Preliminary Trouble Analysis * Fault Localization Clues * Temperature Quick Checks * Transistor Buzz-Out with Voltmeter * Signal-Channel Transistors * Identifying a Stone-Dead Stage * Alignment * Case History * "Hot Chassis" Caution * AGC Checkout and Troubleshooting * Trick of the Trade*

PRELIMINARY APPROACH[1]

Troubleshooting television receivers without service data is dependent upon *previous experience* and the *ability to logically analyze* trouble symptoms and quick-check test data.

Case History

A table-model TV receiver was brought in for repair with a dark screen; sound reproduction was ok. The customer remarked that the picture first started to become defocused and dim with distorted colors. Soon thereafter the screen blacked out. The technician recognized this particular model and had a hunch that the *high-voltage rectifier* had failed. A quick check verified the absence of high voltage, and the technician scheduled the receiver for speedy repair.

When the trouble area must be systematically tracked down, the following steps will provide useful clues. Circumstances alter cases, and the troubleshooter may elect to omit one or more items, depending both on the receiver owner's report, and on the trouble clues that are obvious at the outset.[2]

---

[1]See Chart 7-1.

[2]See Chart 7-2.

CHART 7-1

Functional Sections Associated with Preliminary Trouble Analysis

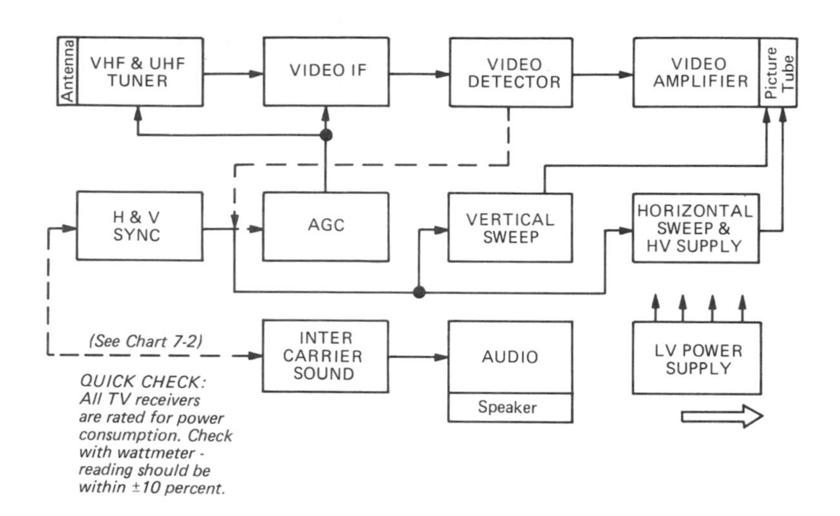

Occasionally, the main functional sections are marked on the under side of the PC board. (Although there is usually some overlap, the markings provide helpful guidelines.)

The troubleshooter keeps 11 functional sections and 3 subsections in mind during his or her preliminary approach, as shown above.

Dotted lines indicate common variations in sectional arrangements, as depicted in greater detail in Chart 7-2. These differences in layout are important, inasmuch as they often affect evaluation of trouble clues in fault localization.

Sectional identification is greatly speeded up by oscilloscope waveform checks (unless the section is completely dead).

Preliminary trouble analysis can also be speeded up in many cases by use of a TV analyzer for high-frequency and low-frequency signal injection, sync-pulse substitution, AGC keying pulse substitution, horizontal and/or vertical drive substitution, high and/or low voltage supply substitution, and related quick checks.

(The rated power consumption of a TV receiver is always indicated on a label secured to the rear cover.)

CHART 7-2

Fault Localization Clues in Relation to Sectional Arrangements

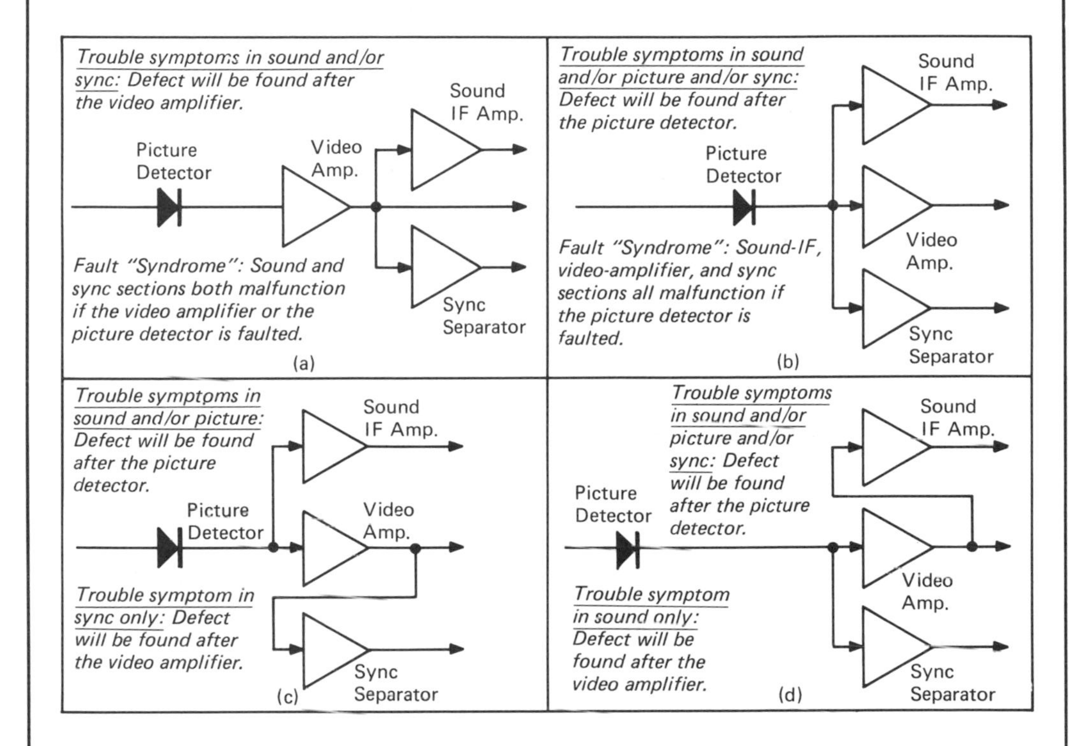

(A) Sound and sync sections following video amplifier; (B) sound, sync, and video sections following picture detector; (C) sound and video sections following picture detector, sync section following video section; (D) video and sync sections following picture detector, sound section following video section.

See Chart 7-4 for explanation of sync trouble symptoms caused by marginal decoupling of receiver sections.

Standard Preliminary Troubleshooting Procedures

Step 1. Observe the picture and sound responses (if any) to variation of the receiver operating and maintenance controls. Look and listen for any changes in picture and/or sound reproduction (or any traces of reproduction, such as sound bars or hum output, "birdies," motorboating, herringbone patterns, noise output, "plops," or "Xmas-tree" patterns.

Step 2. In the case of a blank-raster (no-picture) trouble symptom, observe the snow level on each channel, with the contrast control turned to maximum.[3]

No snow indicates that the video amplifier has no input signal; if there is no sound and little "rushing" from the speaker, start by checking the detector diode. AGC trouble is also a ready suspect.

Medium snow indicates that the IF amplifier has no input signal; if there is no sound and medium "rushing" from the speaker, start by making a tuner-substitution test.

High snow indicates that the trouble does not include the mixer; start by making a local-oscillator signal-substitution test. The RF amplifier is also a ready suspect—try a tuner-substitution test.

Step 3. In the case of a dark screen (no raster), check the high voltage first. If necessary, follow up by measuring the picture-tube terminal voltages.

If all of the picture-tube DC voltages are reasonable, it is indicated that the picture tube is dead. Start by plugging in a picture-tube test jig.

Step 4. If the receiver is stone dead, open the cabinet and try to find some *circumstantial evidence* such as discolored or charred resistors, burned insulation, or puffed-out electrolytic capacitors.

---

[3]From a statistical viewpoint, a no-snow trouble symptom usually indicates a lack of input signal to the video amplifier. However, it can also result from a video-amplifier defect. There is less possibility of a fault in the picture-tube input circuit.

The circuit breaker is usually tripped in this situation. If power can be applied, listen for loud hum from overloaded transformers, snapping arcs; check for hot transistors, resistors, and transformers; look for wisps of smoke or corona discharges; be alert for burning odors (or any odor).

Step 5. Before the troubleshooting job is completed in any event, the chassis should be thoroughly cleaned. If the failure has not turned up in the preceding steps, now is the time to get rid of all the dust, grime, lint, and corrosion that may be present.

This strategy is directed to preparation for a critical visual inspection. Experience proves that the trouble (or important clues) may be spotted before the cleaning job is completed.

Case History: A new table-model receiver was brought in for warranty repair the day following delivery, with a torn-up picture trouble symptom. Visual inspection quickly revealed a small capacitor with one lead broken loose from a poor solder connection. "Spring action" of the other lead caused the disconnected end of the capacitor to tilt up sufficiently that it was easily spotted.

Step 6. Most receivers have one or more plug and connector units. An important part of a visual inspection is checking out these units. (A plug may have been carelessly inserted, for example.)

TEMPERATURE QUICK CHECKS

If you have a temperature probe and DVM, an informative quick check of devices and components can be made, as was explained in Chapter 1. Although temperature quick checks are more meaningful if you have a similar TV receiver in normal working condition for comparison tests, you can nevertheless obtain useful clues in many cases without a comparison receiver.

For example, an electrolytic capacitor "running a fever" at 50°C would be a sure indicator of trouble. Again, a power transistor with "hypothermia," that is operating at 2°C above ambient temperature is almost certainly in big trouble.

TRANSISTOR BUZZ-OUT WITH VOLTMETER

When troubleshooting without service data, follow-up DC voltage measurements are generally made to best advantage by buzzing-out the transistors in sections with malfunctions (or suspected malfunctions), as was explained in Chapter 1. In the case of a stone-dead section, an in-circuit transistor tester can be used to identify the base, emitter, and collector leads of a transistor, as well as its type and its general operating condition.

SIGNAL-CHANNEL TRANSISTORS

In addition to transistor buzz-out with a voltmeter, transistor functions can be determined in most cases by signal-injection tests on the basis of frequency (insofar as signal-channel transistors are concerned). Chart 7-3 shows the basic principles. There are certain "landmarks," also, that are always helpful. For example, the tuner section is easily spotted, and the audio-output circuit is easily spotted, inasmuch as it is connected to the speaker.

IDENTIFYING A STONE-DEAD STAGE

When a stage in the signal channel is stone dead, its operating frequency can usually be determined by checking it with a resonance probe. (The only exception occurs when the fault involves an open coil winding, or when a circuit short happens to shunt the winding.)

The basic resonance probe was depicted in Figure 4-3. To buzz out tuned circuits in TV receivers, the resonance probe is driven by a marker generator or an AM signal generator, as required. For example, your marker generator may not provide 4.5-MHz output; in such a case, an AM signal generator would be used in identification of sound-IF coils or transformers.

As previously noted, the signal level applied to the tuned circuit under test should not be so high that associated transistor junctions are turned on, because the resulting current demand (loading) could confuse the test results. Practically, this means that the probe output voltage should be kept below 150 mV. In most cases, this precaution does not apply because most marker generators and AM signal generators have a maximum output level that limits the probe output

CHART 7-3

Signal-Channel Transistors

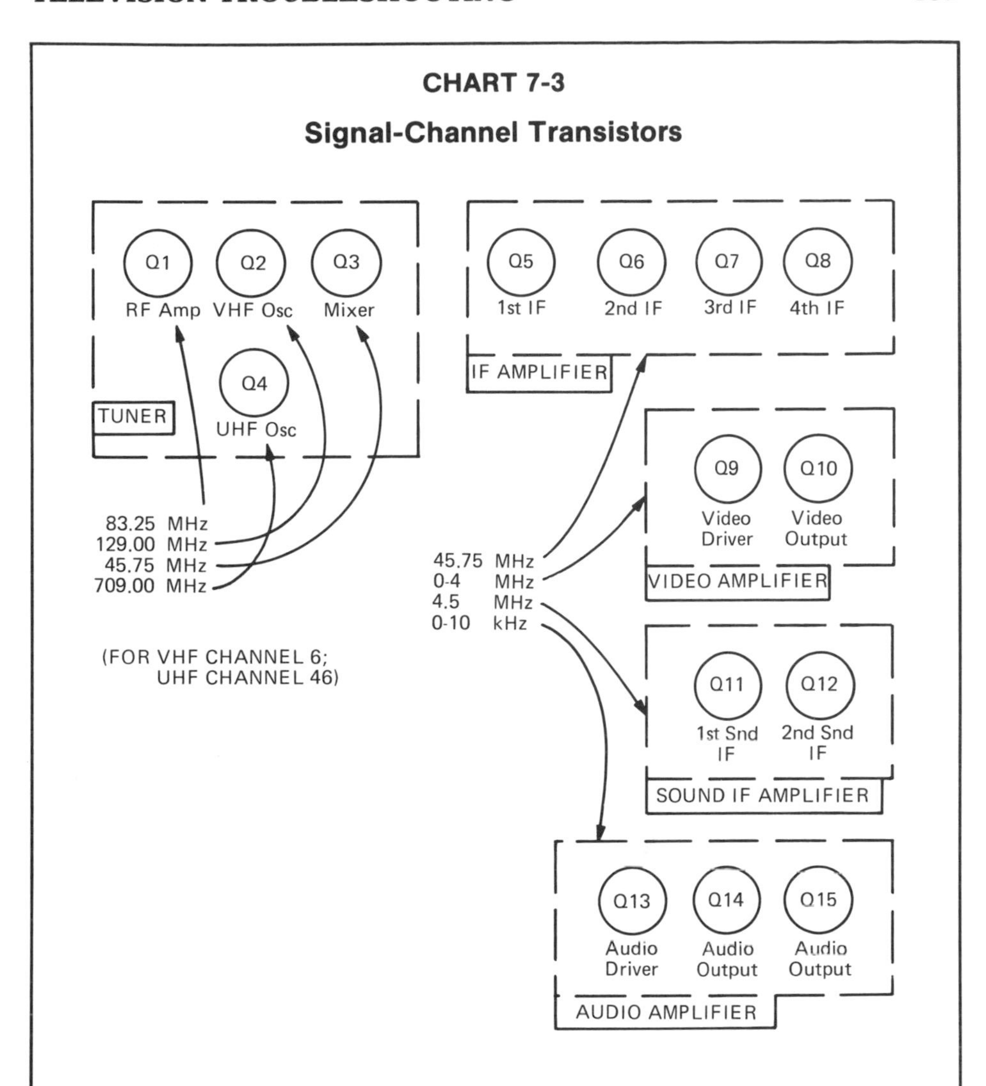

Reasonable collector voltages: RF amplifier, 4 volts; video output amplifier, 90 volts.

Transistors in the signal channel can be identified by signal-injection tests on the basis of frequency (unless the stage or section is completely dead).

A TV analyzer provides suitable test-signal frequencies. Otherwise, a marker generator, AM signal generator, and audio oscillator can be utilized.

The sequence of transistors that operate at the same frequency can be determined on the basis of stage gains, or from the fact that the output signal will be killed when a temporary base-

CHART 7-3 CONTINUED

emitter short-circuit is applied to a transistor following the signal-injection point; on the other hand, the output signal will be unaffected when the short-circuit test is applied to a transistor preceding the signal-injection point.

After a transistor has been buzzed out, it is generally advisable to follow the color-code mapping procedure explained in chapter 3 for audio amplifiers. (The troubleshooter has a considerable list of data to compile, and it is easy to forget the location of a particular device or component unless it is color-code mapped.)

CHART 7-4

Sync Trouble Symptom Caused by Faulty Decoupling and Marginal Filtering (Case History)

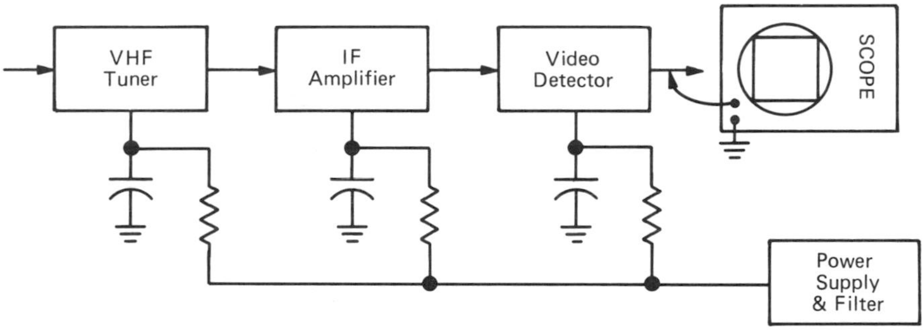

A reasonable ripple voltage on the V_{CC} line is 15 mV, r.m.s., or 42 mV, p-p.

The trouble symptom in this example was intermittent "picture rolling" when the receiver was tuned to a network program. In other words, the picture rolled vertically for a short time, and then locked in sync for a longer or shorter time after which it would again roll vertically for a short time. However, the trouble symptom did not occur when the receiver was tuned to a local TV station.

CHART 7-4 CONTINUED

This distinction between network reception and local-station reception provided the clue that the troubleshooter needed to run down the fault.

During local-station reception, the receiver and the TV transmitter are powered from the same public utility. On the other hand, during network reception, the receiver and the TV transmitter are powered from different public utilities (generating stations). The practical result is that there is a small and unpredictable difference in the 60-Hz power frequencies at the receiver and at the network transmitter. In turn, there is a slow beat between the vertical sync pulse in the network video signal and the power-supply hum voltage at the receiver.

When an oscilloscope was connected at the picture-detector output, and the receiver was tuned to a network program, the troubleshooter observed that the vertical-sync pulse would occasionally die out at the same time that the picture started to roll vertically. Then, the vertical-sync pulse would come in again, whereupon the picture locked in vertical sync.

Next, an oscilloscope check of the V_{CC} line showed an abnormal "ripple and hash" component on the line. The result was that the incoming signal beat against this spurious V_{CC} component voltage in the tuner and IF amplifier sections. Investigation then showed that although the individual RC decoupling circuits were not defective, the filter capacitor (which also serves a decoupling function) had lost a large portion of its capacitance.

Bottom line: When the troubleshooter replaced the output filter capacitor, the intermittent picture rolling stopped.

voltage to approximately 100 mV in this test. (See Figure 7-1 for a TV IF resonance-probe arrangement.)

Remember that it can be helpful to know the typical signal-voltage levels in normal operation. These levels are indicated in Figure 7-2. The two most important reference levels are the picture-detector output voltage and the video-amplifier output voltage. The picture-detector output voltage can be checked either with a DC voltmeter (inasmuch as the detector provides rectifier action), or with a calibrated oscilloscope. The video-amplifier output voltage must be checked with an oscilloscope.

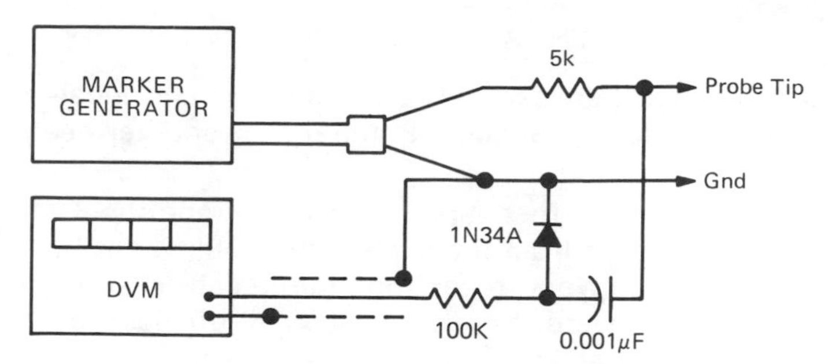

To identify TV IF transformers, the marker generator is set for a DVM indication of approximately 100 mV with a test frequency of about 44 MHz. When the probe tip and ground clip are applied across a tuned coil, the DVM reading will drop to a much lower value. However, as the marker generator is tuned to the peak resonant frequency of the coil, the DVM reading will rise to a maximum. Then, as the marker generator is tuned past the resonant frequency of the coil, the DVM reading will start to fall.

Note that in some cases, when the resonance probe is applied across a high-Q coil, voltage magnification may be observed at the peak resonant frequency. In other words, the DVM reading will be greater than when the probe tip is disconnected from the coil.

When a comparatively low-Q coil is under test, no voltage magnification will be observed as the peak resonant frequency, and the DVM reading will be somewhat less than when the probe tip is disconnected from the coil. (However, the DVM reading will rise to a maximum as the generator is tuned to the peak resonant frequency of the coil, whether its Q value is low or high.)

Sometimes a stone-dead stage has a tuned-circuit fault; for example, a coil may be shunted by a short-circuited capacitor. In such a case, a resonance probe shows zero response. The troubleshooter must fall back on landmarks and ohmmeter tests.

Figure 7-1 A TV IF resonance-probe arrangement.

ALIGNMENT

In general, alignment should not be started until after all troubleshooting has been completed. However, *there is one exception:* if it is known that the operator of the receiver has tampered with the alignment adjustments, and if it is found that the IF amplifier is oscillating, a preliminary alignment procedure should be followed at

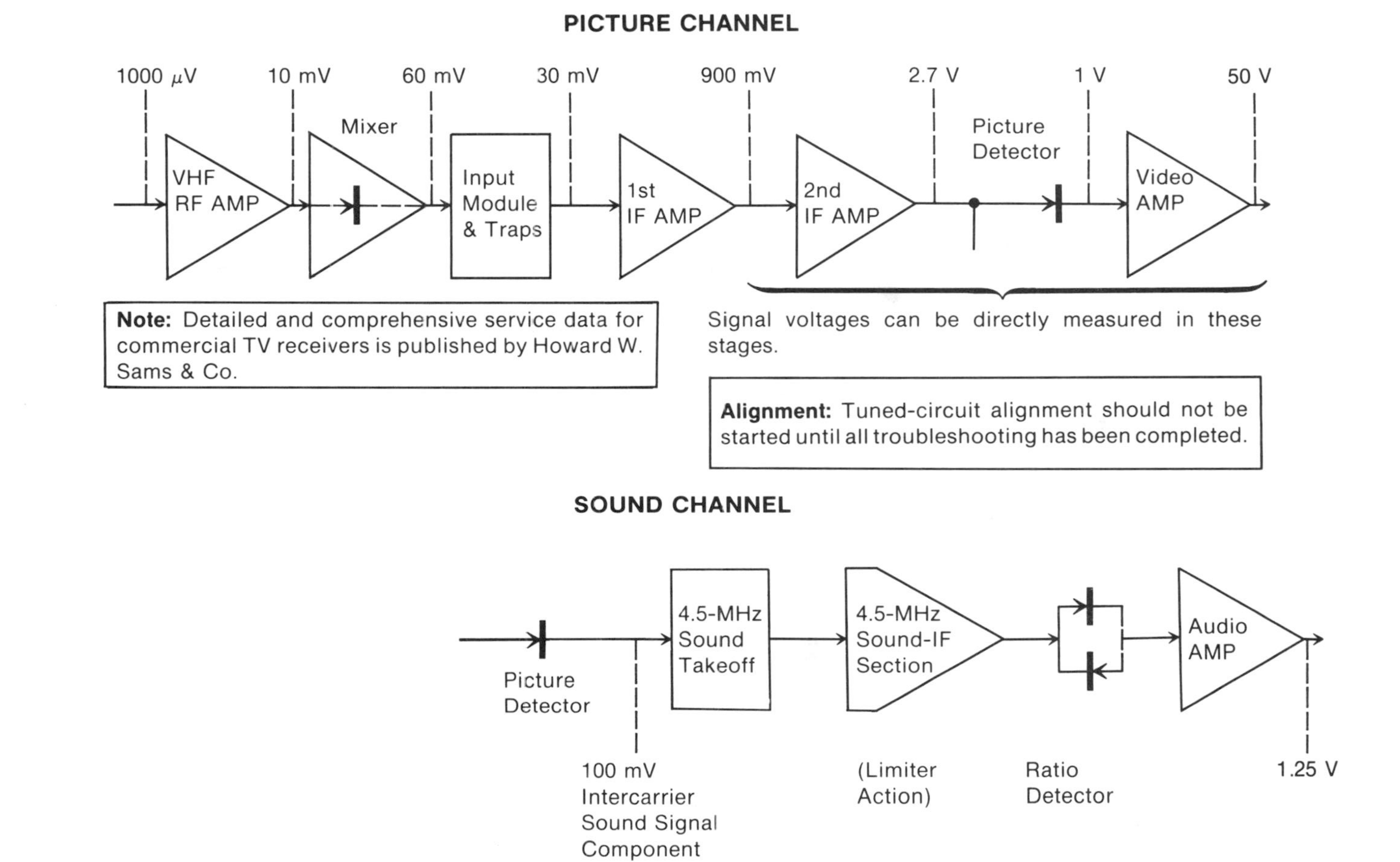

(The most practical way of measuring high-frequency signal-voltage levels is to inject a test signal from a calibrated generator and adjust its level to obtain specified output voltage from the video amplifier.)

Representative VHF, IF, VF, and AF voltage levels.

Figure 7-2 Typical signal-voltage levels in a small black-and-white TV receiver.

the outset. (In some cases, this may be all that is required to restore the receiver to normal operating condition.)

An IF amplifier may break into uncontrollable oscillation if primaries and secondaries in the IF transformers are peaked up to the same frequency. When the IF amplifier takes off, the picture-tube screen generally displays only a raster with little if any snow. A DC-voltage check at the output of the picture detector will show an abnormally high voltage, in case the IF amplifier is oscillating.

When the alignment adjustments have not been tampered with, and the IF amplifier is oscillating, the trouble will almost always be tracked down to open bypass or decoupling capacitors. A check with a peak-reading probe and DVM might show, for example, that the AGC line is hot, due to one or more open capacitors. In such a case, positive feedback of IF signal voltages from an IF stage to a preceding IF stage can cause self-oscillation.

Case History

A TV receiver was brought in for service with the complaint that reception could be obtained only when the channel-selector switch was turned counterclockwise. If the channel-selector switch was turned clockwise, the picture-tube went blank, displaying a raster and practically no snow. Suspecting that this was an unusual type of IF self-oscillation, the troubleshooter bridged a 0.1 μF capacitor from the AGC line to chassis ground. In turn, normal operation resumed, verifying the preliminary diagnosis.

"Hot Chassis" Caution

Some TV receivers have transformerless power supplies, with one side of the 117-volt line connected to the receiver chassis, or common ground. This type of power supply may give the unwary technician an electric shock, and may also damage test equipment to which it is connected.

To avoid this hazard, it is good practice to power the receiver from an isolation transformer. It is true that danger can be avoided by inserting the power plug into the outlet so that the grounded side of the line is connected to the chassis ground in the receiver.

However, the problem here is that a technician tends to become distracted, and to forget to check which way the power plug is inserted in the outlet. Therefore, it is good practice always to power the receiver under test from an isolation transformer.

AGC CHECKOUT AND TROUBLESHOOTING

When troubleshooting without service data, it is sometimes essential to keep the basic types of AGC arrangements clearly in mind. As previously noted under radio circuitry, either forward-AGC or reverse-AGC may be employed. *In some TV receivers, both forward- and reverse-AGC are utilized.* Amplified AGC is often used, with direct coupling of AGC voltage through successive IF stages.

Most TV receivers have *delayed-AGC* circuitry whereby the gain of the tuner is maintained at maximum while the IF gain is initially reduced. Then, as the IF gain is further reduced, the tuner gain is progressively reduced. Delayed AGC improves weak-signal reception. All but the simplest types of receivers include *keyed-AGC* circuitry to obtain improved noise immunity.

RF AGC

The RF-amplifier gain in all TV receivers is varied in accordance with the incoming signal strength by AGC action which changes the base-emitter bias on the RF-amplifier transistor. Gain control in the case of reverse AGC is based on the drop of transistor beta at low values of emitter current. Gain control in the case of forward AGC is based on the drop of transistor beta at saturation values of emitter current.

Note in passing that the transistor's input and output resistances also change considerably over the AGC operating range, with the result that low gain is accompanied by mismatch to source and load impedances. Mismatch further reduces the signal-power gain of the stage. Note also that AGC action may be designed to reduce the stage gain to less than unity (to provide an insertion loss) when the incoming signal strength is very high.

AGC bias voltage is not necessarily applied to the base of the RF transistor in all receivers—the AGC bias voltage may be applied to the emitter. Of course, AGC voltage applied to the base has reverse polarity with respect to AGC voltage applied to the emitter of an RF transistor.

It is worth remembering that even when the AGC voltage is so high that the RF transistor is completely cut off, the insertion loss of the RF transistor is not infinite—there is normally a small residual feedthrough signal. This low-level signal output is the result of transistor junction capacitances and the fact that complete neutralization is not utilized in commercial TV receivers.

IF AGC

With reference to Figure 7-3, a widely used IF AGC arrangement is shown, which the troubleshooter should keep in mind. Although it *appears* that AGC voltage is applied to only the first IF stage, the direct coupling of the stages results in AGC voltage being applied to the second and third stages also.

Note in Figure 7-3 that forward AGC is utilized in the first two IF stages, whereas reverse AGC is applied to the third IF stage. The configuration functions as follows:

1. In the absence of an incoming signal, the AGC terminal rests at a positive reference voltage of approximately 2.1 volts in normal operation.
2. Since the emitter of the first IF transistor is returned through a resistor to a +12 volt supply, the base of the first transistor is negative with respect to the emitter, and the base-emitter junction is forward-biased.
3. Conduction in the first IF transistor causes the emitter voltage to rest at about 2.4 volts, so that a forward-bias voltage of 0.3 volt (2.4 − 2.1 volts) is normally measured between base and emitter. (This is an example of germanium transistor design.)
4. This conduction in the first IF transistor also causes a voltage drop of 9.6 volts across the emitter resistor, and a collector-to-emitter voltage drop of 2.4 volts.
5. Next, as the incoming signal strength starts to increase from zero, the voltage applied to the AGC terminal swings less positive. (This is in consequence of the fact that the RF-amplifier transistor in this design does double duty as an AGC amplifier, and its AGC-voltage output is less with signal present.)
6. Since the AGC terminal becomes less positive, the forward bias on the first IF transistor increases. Accordingly, the voltage drop across its emitter resistor also increases. The resulting increased conduction in the first IF transistor provides forward-AGC action for the second IF transistor. (The first IF transistor also provides "delayed-AGC" action.)

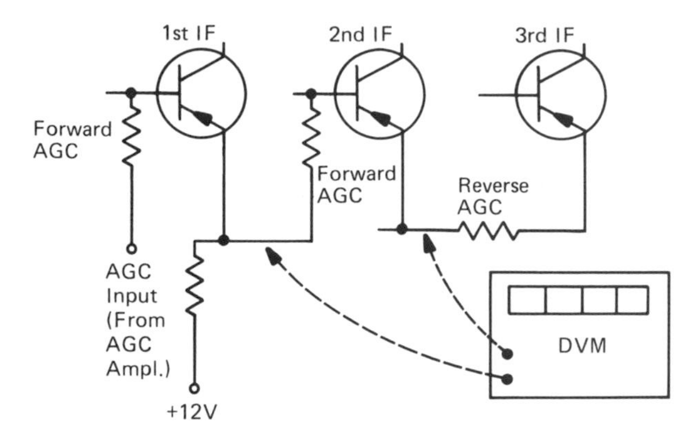

Note: When AGC trouble is suspected, preliminary troubleshooting should start by clamping the AGC line from a bias box. When troubleshooting without service data, the normal operating voltage is unknown. However, the AGC line will be clamped at the correct value when the IF transistors have normal base-emitter bias voltage (about 0.6 volt for silicon transistors). When the first IF transistor has been provided with a normal base-emitter bias voltage, the second and third IF transistors should also have approximately the same base-emitter voltages. If DC-voltage measurements show that there is a substantial difference in base-emitter bias voltages of the three transistors, look for a defective transistor or for a component defect such as a leaky AGC capacitor.

The origin of the AGC voltage is typically in an emitter-follower stage following the picture-detector diode.

Figure 7-3 A representative IF AGC network; the AGC voltage is passed along through the 1st IF and 2nd IF to the 3rd IF transistor.

7. This forward-AGC action causes a reduction in collector-to-emitter voltage, and at values below one volt, the stage gain diminishes rapidly. In other words, as the collector voltage varies between one and two volts, there is little change in stage gain. (This is a built-in AGC delay action.)
8. The emitter voltage with respect to ground in the first IF stage drops from a maximum of +2.4 volts toward zero as more AGC voltage is applied. This emitter voltage drop provides AGC bias for the second IF transistor. The AGC bias is applied through a resistor to the base of the second IF transistor.
9. The second IF transistor is AGC-controlled in the same

manner as the first IF transistor. This is just another way of saying that an increase in signal strength causes the base voltage of the second IF transistor to drop from a maximum of 2.4 volts to some lower value.

10. As explained above for the first stage, the emitter voltage in the second stage is applied as AGC bias to the third IF transistor. In this case, however, the second stage affects the emitter voltage (not the base voltage) of the third IF transistor. Stated otherwise, the third IF stage operates with reverse AGC.
11. The base voltage of the third IF transistor is initially held at a fixed value by a resistive voltage divider. Under no-signal conditions there is no AGC voltage, and the emitter rests slightly positive with respect to the base. (This resting bias normally provides maximum gain in the third IF stage.)
12. Next, as the incoming signal level increases from zero, the emitter voltage of the second IF transistor and the emitter voltage of the third IF transistor become less positive. (The emitter of the third IF transistor becomes less positive with respect to its base, and the forward-bias voltage decreases.)
13. The result is that reverse AGC is employed in the third IF stage. The reason for using the foregoing combination of forward-AGC and reverse-AGC is to stabilize the IF bandwidth over a wide range of signal levels. (In other words, forward AGC tends to reduce bandwidth, whereas reverse AGC tends to increase bandwidth. In turn, these opposing actions tend to cancel out, and to maintain a constant IF bandwidth.)

COMPARISON QUICK CHECK

When AGC trouble is suspected and a similar receiver is available for comparison tests, a helpful quick check can be made by measuring the resistance from the AGC line to ground with a low-power ohmmeter. If there is a substantial discrepancy in the resistance readings for the two receivers, the troubleshooter concludes that the suspicion is verified. An off-value resistance reading is commonly tracked down to a leaky AGC capacitor.

However, other faults can also be responsible, such as defective transistors or diodes.

Spotting AGC Transistors

Even when a similar receiver is available for comparison checks, the troubleshooter must buzz out and identify the AGC transistors when AGC trouble is suspected. From one to three AGC transistors may be utilized. AGC controls are associated with the AGC transistors. For example, the troubleshooter expects to spot an AGC control and a bias delay control, as explained in greater detail in Chapter 8. Sometimes, apparent AGC trouble is tracked down to incorrect adjustments of the AGC controls.

The most important clue to spotting AGC transistors is the change in DC voltage at the collector of an AGC transistor when the signal level is changed. A signal generator is not required for a quick check—you can tune the receiver to a strong station signal and the short-circuit the antenna-input terminals to check whether a change occurs in DC collector voltage.

As detailed in Chapter 8, RF and IF transistors in the signal channel may do double duty, functioning both as high-frequency amplifiers and as DC voltage amplifiers in the AGC "loop." To distinguish between a signal-channel transistor and an AGC transistor in case of doubt, make a follow-up quick check with a resonance probe (Figure 7-1).

If an AGC transistor is tested with a resonance probe, it will look like a very low impedance load over the 40-MHz range. On the other hand, an IF transistor that looks like an AGC transistor will peak up unmistakably when quick checked with a resonance probe.

TRICK OF THE TRADE

Intercarrier-IF stages can be buzzed out with a resonance probe if the power is turned off, or if the intercarrier section is dead. On the other hand, if there is sound output from the speaker, intercarrier-IF stages can be spotted by employing a trick of the trade, as shown in Figure 7-4.

This is an aural signal-tracing method in which the 4.5-MHz signal in the intercarrier-IF section is heterodyned with the output from a

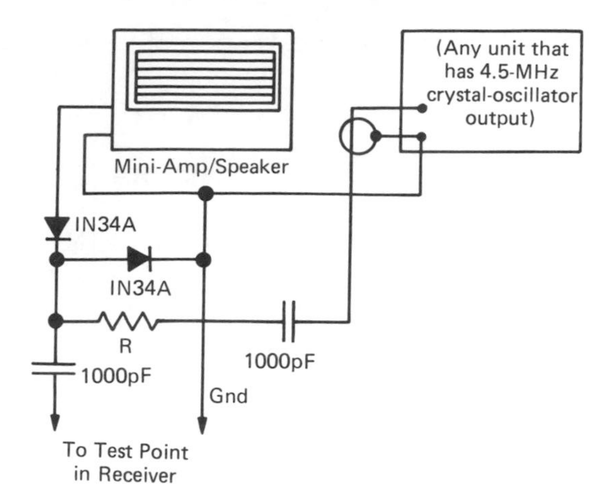

This is a beat type of aural signal tracer. The 4.5-MHz output voltage from the crystal oscillator heterodynes with an intercarrier-IF signal. In turn, an audio tone is heard from the speaker.

The value of the mixer resistor R is chosen to provide optimum sound output from the speaker. Also, the 4.5-MHz output voltage from the crystal oscillator should be adjusted for best sensitivity.

The pitch of the beat tone depends on the precise frequency difference between the intercarrier-IF signal and the 4.5-MHz crystal oscillator.

The miniamp/speaker may be a Radio Shack No. 277-1008. To determine the best value for R, start with a 1-kilohm value, and try higher and lower values.

Note that an AM signal generator can be used, if desired, instead of a crystal oscillator. However, a signal generator must be carefully tuned—service generators may also tend to drift in frequency, particularly until they have warmed up.

Figure 7-4 Aural signal tracer for intercarrier-IF circuitry.

4.5-MHz crystal oscillator. In turn, if there is an intercarrier-IF signal in the circuit under test, a beat tone will be heard. A beat tone is generated because it is highly improbable that the output from a 4.5-MHz crystal oscillator will be *exactly* the same as the intercarrier-IF frequency.

SHORT-WAVE RADIO RECEIVER SERVES AS INTERCARRIER SIGNAL TRACER

Another useful trick of the trade is shown in Figure 7-5. This

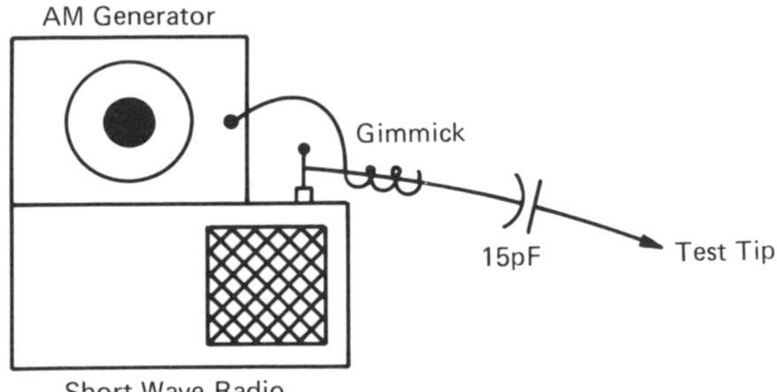

This intercarrier-IF signal tracer utilizes a small short-wave radio receiver instead of the miniamp/speaker and external modulator arrangement depicted in Figure 7-4.

The output from the AM generator is mixed with the signal under test by means of a gimmick (several turns of wire that provide capacitive coupling to the rod antenna of the radio receiver).

A test lead is terminated by a 15-pF capacitor for probing in the circuitry under test.

The short-wave radio receiver is tuned to 4.5 MHz, and the AM generator is set to a slightly higher or a slightly lower frequency.

When the test tip is applied in an intercarrier-IF circuit that has a 4.5-MHz signal present, a beat-frequency tone will be audible from the speaker.

This signal tracer enables the troubleshooter to quickly determine where an intercarrier-IF signal is being stopped (or substantially attenuated) in the intercarrier-sound channel.

Figure 7-5 A short-wave receiver can be utilized as an intercarrier signal tracer.

arrangement employs a short-wave radio receiver and an AM signal generator as a 4.5-MHz intercarrier-sound IF signal tracer. It is basically quite similar to the signal tracer depicted in Figure 7-4, inasmuch as it develops an audio beat tone when the test tip is applied in a circuit that has a 4.5-MHz signal present.

8

Additional Television Troubleshooting Methods

*Closed-Loop AGC * Video-Amplifier Troubleshooting * Buzzing Out Video Amplifier Transistors * Intercarrier Sound Section * Sync-Circuit Troubleshooting * Buzzing Out Sync-Section Transistors * Widely Used Sync Circuitry * TV Troubleshooting with the Oscilloscope * Using Dual-Trace Oscilloscopes * Waveform Tolerances*

CLOSED-LOOP AGC

Troubleshooting without service data is greatly facilitated by a clear understanding of the basic AGC arrangements. One of the most popular configurations is a closed loop consisting of the AGC gate, the RF amplifier, IF amplifiers, video detector, and first video amplifier, as exemplified in Figure 8-1.[1]

This AGC system normally maintains a relatively constant 1.2-volt output at the emitter of the first video amplifier over a wide range of signal-input voltage. It is a basic gated or keyed AGC arrangement, wherein an AGC voltage is "pulsed" into the AGC capacitors at the horizontal-sync rate; capacitive storage maintains the prevailing AGC level during scan time from one horizontal sync pulse to the next.

A keyed-AGC system is normally immune to most noise voltages and is completely unaffected by the camera-signal waveform. This immunity to interference results from the fact that the video signal is sampled only for the duration of the horizontal sync tips. Operation of the keyed-AGC configuration is as follows:

---

[1]Any AGC arrangement can be clamped during troubleshooting procedures. The clamp voltage is correct when there is slightly over 0.6 volt forward bias on the IF transistors.

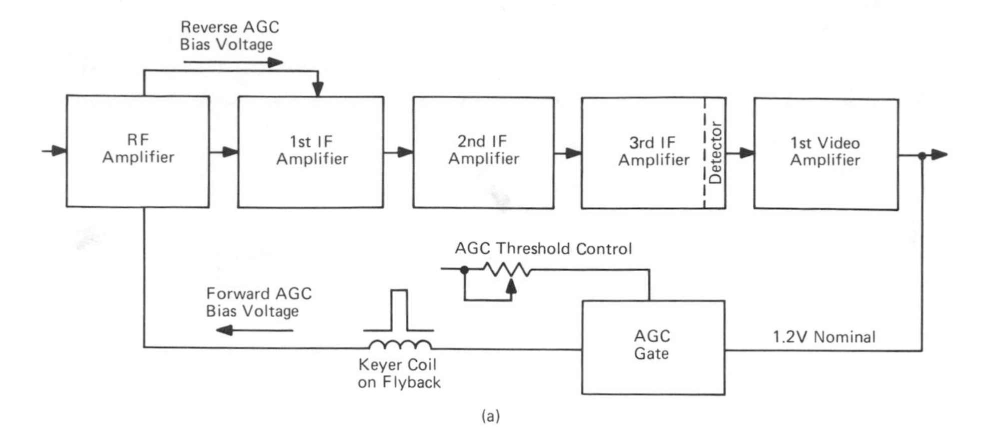

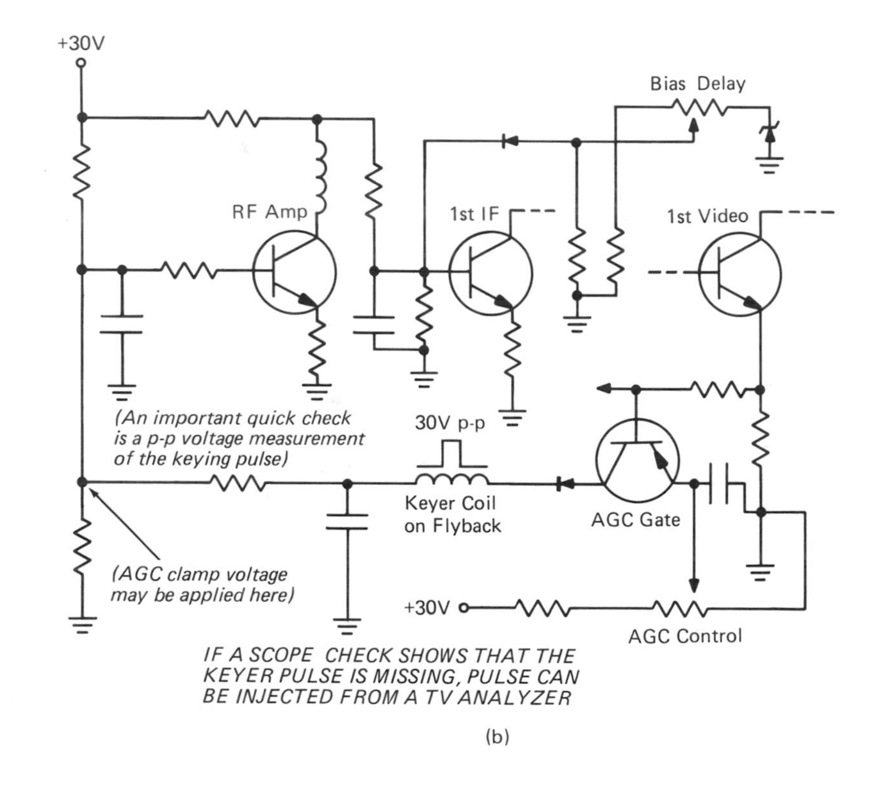

Figure 8-1 A widely used keyed-AGC arrangement. (a) Block diagram; (b) skeleton schematic.

1. As the input signal to the RF amplifier increases, the IF and video signal level tends to increase, and the output from the first video amplifier tends to increase.
2. However, this tendency is largely cancelled, inasmuch as the input signal to the AGC gate starts to increase also. The AGC gate is keyed into conduction *for the duration of the horizontal sync tip* by a 30-volt negative pulse from the flyback transformer.
3. This keyer pulse is applied as V_{CC} to the collector of the AGC gate transistor, which in turn amplifies the horizontal-sync voltage present at its base.
4. Normally, a positive AGC voltage is produced at the collector of the AGC gate; *this AGC voltage is retained during horizontal scan time by the long time constant of the AGC bus.*
5. Note that in order to prevent the collector-base junction of the AGC gate transistor from becoming forward-biased by this positive AGC voltage, a diode is connected in series from the collector to the AGC bus. If this diode becomes leaky or short-circuited, the time-constant of the AGC bus is then incapable of retaining the AGC voltage value during horizontal scan time. A typical trouble symptom is "shaded raster" from left to right on the picture-tube screen.
6. The positive AGC voltage produced at the collector of the AGC gate transistor is applied as forward bias voltage to the RF amplifier transistor, thereby reducing its gain as the signal level increases.
7. The RF amplifier in this arrangement does double duty; *it amplifies the RF signal, and it also functions as a DC amplifier for the AGC voltage.* It also reverses the polarity of the AGC voltage due to CE mode operation, and in turn applies reverse AGC bias voltage to the base of the first IF amplifier transistor.
8. In normal operation, the RF amplifier provides nearly full gain on weak-to-medium signals, with most of the AGC action occurring in the IF amplifier. As the signal level further increases, the RF amplifier provides increasing RF gain reduction.
9. Approximately 35 dB of signal attenuation is normally possible by means of AGC action on the first IF transistor,

and an additional 35 dB attenuation is provided on strong signals in the RF amplifier transistor.

10. Control of the level at which AGC action becomes effective in the RF amplifier transistor is provided by the "bias delay" control in Figure 8-1. This control establishes the minimum gain of the first IF amplifier transistor.
11. The AGC voltage applied to the RF stage passes through an RC filter circuit; its time constant is sufficiently slow to prevent "hang-up" of sync or AGC in case the horizontal scan is out of sync lock, but its time constant is sufficiently fast to permit adequate AGC action on fast signal fluctuations. If "airplane flutter" or other trouble symptoms related to AGC action occur, *the troubleshooter should check out the AGC filter circuit.* For example, resistors sometimes increase in value and change the time constant of the AGC bus. On the other hand, a decrease in value of a shunt resistor results in shaded raster from left to right on the picture-tube screen. Similarly, a leaky AGC capacitor can cause a shaded raster trouble symptom.

VIDEO-AMPLIFIER TROUBLESHOOTING

As shown in Figure 8-2, a typical video amplifier comprises an emitter follower driving a common-emitter output stage. Note that the output circuit of the video detector is also the input circuit of the first video amplifier. Trouble symptoms in the video amplifier usually show up in the picture display. However, as previously noted, the AGC voltage may have its source in the first video amplifier, and symptoms of AGC trouble may actually be video-amplifier malfunctions.

Note also in Figure 8-2 that the intercarrier sound channel has its source (in this example) at the collector of the first video-amplifier transistor. In turn, symptoms of intercarrier-sound trouble may actually be in the first video-amplifier stage. As a rule of thumb, any video-amplifier malfunction that impairs sound reproduction will also impair picture reproduction. (This "split" cannot be made in preliminary analysis of AGC trouble, inasmuch as AGC malfunction impairs picture reproduction regardless of its source.)

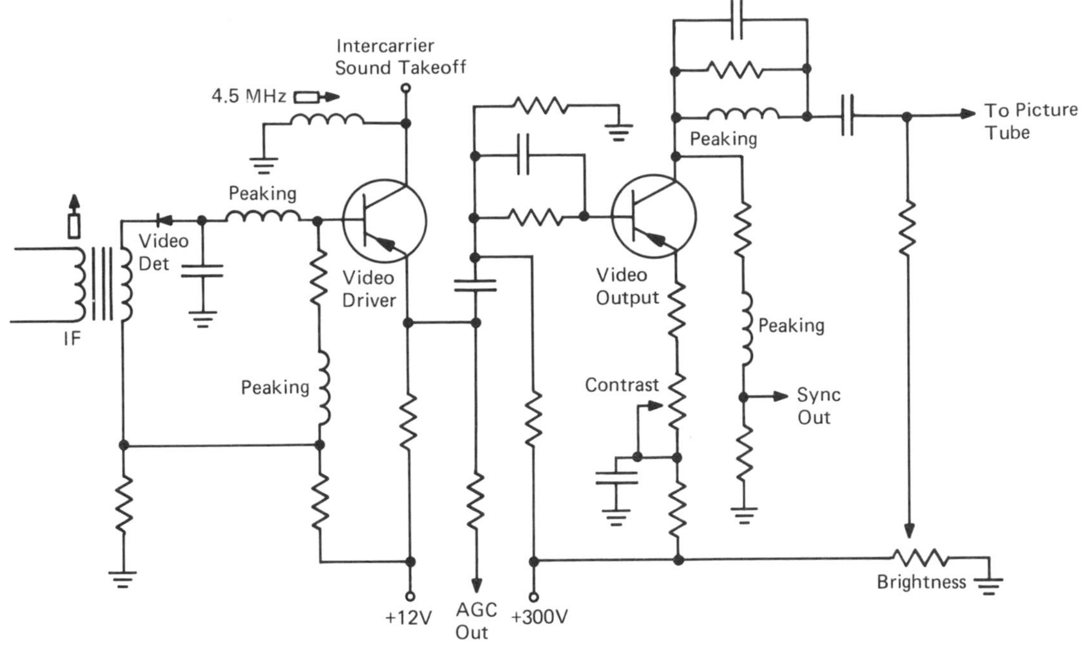

When troubleshooting without service data, the video-output transistor can be "spotted" with a DC voltmeter because it operates with a higher collector voltage than other transistors.

In this example, series and shunt peaking coils are utilized in both the driver and output stages. The peaking coils are sometimes slug-adjustable, so that the technician can optimize the amplifier frequency response. Both transistors normally operate in class A. Incorrect bias voltage results in impaired picture reproduction. Intercarrier buzz can be produced by incorrect bias on the driver transistor. Bias shift is often tracked down to leaky capacitors.

Figure 8-2 A basic video-amplifier configuration.

Video-amplifier operation is checked to best advantage with an oscilloscope. A signal input of 1 volt p-p is normally stepped up to 50-100 volts p-p at the picture-tube input (with the contrast control set to approximately midrange). Picture distortion is associated with poor frequency response. A video amplifier normally provides a uniform output voltage from 60 Hz to 3.5 MHz, or sometimes 4 MHz. Frequency response can be easily checked with a wide-range audio generator and a DVM operated on its AC-voltage function.

The practical value of a video-frequency sweep generator and oscilloscope in video-amplifier checkouts is often overlooked. In other words, a video-frequency sweep generator shows the gain and complete frequency response of the video amplifier at a glance. The generator is used to drive the video amplifier, and the amplifier output is either fed directly to a wide-band scope, or the amplifier output is passed through a demodulator probe and thence to the scope.

A VF sweep generator has a maximum output voltage, such as 0.5 volt p-p. In turn, a calibrated scope connected to the output of the video amplifier indicates the amplifier gain. For example, if the scope pattern has an amplitude of 30 volts p-p, the troubleshooter recognizes that the video-amplifier gain is 60 times.

Buzzing Out Video-Amplifier Transistors

Buzzing out video-amplifier transistors is comparatively easy, provided that at least some video-signal display is visible on the picture-tube screen. An audio oscillator is a suitable signal source; when an AF voltage is injected via a blocking capacitor into the base of a video-amplifier transistor, "bars" are displayed on the picture-tube screen.

On the other hand, if the AF voltage is injected into the base of an IF transistor, for example, no picture-tube display is obtained. Similarly, the AGC transistors do not give a visible response to an injected AF voltage.

When there is a "dark-screen" trouble symptom, or a "raster only" trouble symptom to start with, other quick checks must be used to buzz out the video-amplifier transistors. As noted in Figure 8-2, the video-output transistor can be identified on the basis of DC-voltage measurements, inasmuch as its collector operates at a higher voltage than other transistors in the receiver.

Remember that a video-frequency sweep generator and scope are also very handy for buzzing out the video-amplifier transistors in a receiver when the test signal is not reproduced on the picture-tube screen.

Low gain in a video amplifier is commonly caused by defective capacitors; transistors can become leaky, open, or shorted. Sometimes off-value resistors cause video-amplifier trouble symptoms. (Remember that the video-detector diode operates in the video-amplifier input circuit—don't forget to check the diode.)

As a practical observation, it is good practice always to use a low-capacitance probe with the scope when checking video-amplifier circuitry. The probe minimizes circuit loading so that false conclusions concerning frequency response (in particular) are avoided. Another practical pointer in preliminary troubleshooting procedures is to connect a 10μF blocking capacitor in series with the "hot" output lead from the sweep generator (or other generator).[2] This precaution will prevent accidental short-circuits and possible damage to transistors when checking unfamiliar circuits.

INTERCARRIER SOUND SECTION

The intercarrier-sound section of a TV receiver is similar in basic principles to the IF section in an FM radio receiver, as explained in previous chapters. The essential difference between them is that an intercarrier-IF amplifier has a center frequency of 4.5 MHz, whereas a radio FM-IF section has a center frequency of 10.7 MHz.

Another distinction is in bandwidths; an intercarrier-IF amplifier has a rated bandwidth of 50 kHz, whereas a radio FM-IF section has a rated bandwidth of 150 kHz. Ratio detectors are commonly used for FM detection, and the intercarrier-IF section usually provides some limiting action to assist in rejection of amplitude modulation.

Buzzing Out Intercarrier-IF Transistors

When troubleshooting without service data, the first requirement in checking out the sound section is to identify the intercarrier-IF transistors. If the audio section and speaker are workable, the job is easy; all that is necessary is to inject a 4.5-MHz signal from an AM generator into the transistor under test. If the 4.5-MHz signal is modulated from 50 to 100 percent, it will have appreciable incidental frequency modulation, and the test signal will be audible through the limiter and the ratio detector.

In situations where the audio section is dead, other tests must be used to buzz out the intercarrier-IF transistors. The easiest way, in most cases, is to use a resonance probe to spot the 4.5-MHz resonant circuits associated with the intercarrier-IF transistors. A typical TV receiver employs two intercarrier-IF transistors with three 4.5-MHz tuned circuits. Receivers that utilize integrated sound circuitry typically include two 4.5-MHz tuned circuits.

---

[2]Blocking capacitors used in preliminary troubleshooting procedures should be of nonpolarized design and rated for comparatively high working voltage.

Intercarrier buzz, also called sync buzz or buzz in the sound, is a harsh 60-Hz interference that usually changes intensity with changes in picture-background brightness. It stems from the vertical sync pulse, and results from cross-modulation of the picture signal with the sound signal.

Cross-modulation causes the FM sound signal to become amplitude-modulated by the sync pulses; the vertical sync pulse is within the audio-frequency range. Although residual cross-modulation is always present, intercarrier buzz is normally inaudible, for the following reasons:

1. The RF and IF sections are in proper alignment, so that the level of the sound carrier does not exceed 10 percent of the peak response in the IF channel.
2. IF transistors are operating in a reasonably linear mode, without peak overload.
3. The first video-amplifier transistor is operating in a reasonably linear mode (sound takeoff is commonly located in the first video-amplifier stage).
4. Normal limiting action is taking place in the intercarrier sound section. (See Figure 8-3.)
5. Ratio-detector diodes are well matched, the stabilizing capacitor is not leaky or open, and the tuned circuits are properly aligned.

SYNC-CIRCUIT TROUBLESHOOTING

When troubleshooting sync circuitry, it is advantageous to make comparison tests on a similar receiver that is in good working condition. If a comparison receiver is not at hand, the troubleshooter must proceed by dead reckoning. This requires a clear understanding of sync-circuit functions and trouble-symptom analysis.

Buzzing Out Sync Section Transistors

The oscilloscope is the most important tool for buzzing out sync-section transistors. Most troubleshooters are familiar with the appearance of the complete video signal, stripped sync tips, horizontal pulse trains, differentiated pulse trains, vertical pulse

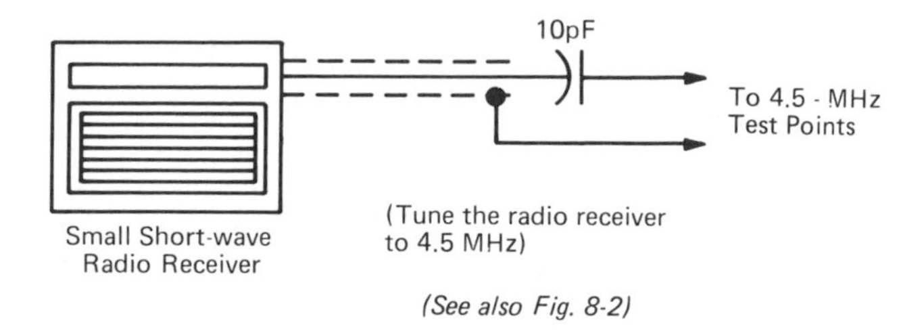

If a wide-band oscilloscope is not available, you can use a small short-wave radio receiver to signal-trace sync-buzz interference through the 4.5-MHz intercarrier circuitry.

Sync-buzz interference amplitude-modulates the 4.5-MHz carrier and produces a harsh 60-Hz rasp in the speaker.

Signal-tracing intercarrier sound sync buzz is most informative when made on a comparative basis, with respect to a similar TV receiver that is in normal operating condition. However, useful comparative tests can also be made with respect to a somewhat similar TV receiver, inasmuch as intercarrier sound circuitry is fairly standardized.

Note that the signal tracer will pick out the intercarrier-sound signal at the output of the video detector, and through the video-amplifier section. In turn, the troubleshooter can easily determine whether the sync buzz is being generated in the IF section or in the video-amplifier section.

Sync buzz should not be confused with vertical-sweep buzz. In other words, you will sometimes discover that spurious cross-coupling between the vertical-sweep section and the sound section is causing 60-Hz buzz.

To distinguish between sync buzz and sweep buzz, roll the picture by turning the vertical-hold control. No change in pitch will result in the case of sync buzz. On the other hand, a change in pitch will be heard in the case of sweep buzz.

Figure 8-3 Sync buzz can be traced in the intercarrier circuitry with a small short-wave radio receiver.

trains, and integrated pulse trains. Even when a stage is completely dead, the drive waveform at the input to the stage will usually clue in the troubleshooter regarding stage function.

Three jobs are normally performed by the sync section:

1. Sync pulses (tips) are separated from the video signal, and the video signal is rejected.
2. Stripped sync pulses are "squared up," and are limited to a fixed amplitude.

3. Vertical pulses are separated from horizontal pulses by means of integrating and differentiating circuits.

If there is trouble in the signal channel, the input signal to the sync section may be objectionably distorted, or missing. In such a case, the troubleshooter may use a TV analyzer to inject a normal input signal to the sync section.

Basic sync circuitry is exemplified in Chart 8-1. It follows from the description of circuit action in the chart that failure of the sync separator to provide good clipping action points to *capacitor leakage*, or to *transistor junction leakage*. Off-value resistors can also cause poor clipping action, although this fault is relatively infrequent. (See Chart 8-2.)

CHART 8-1

Widely Used Sync Circuitry

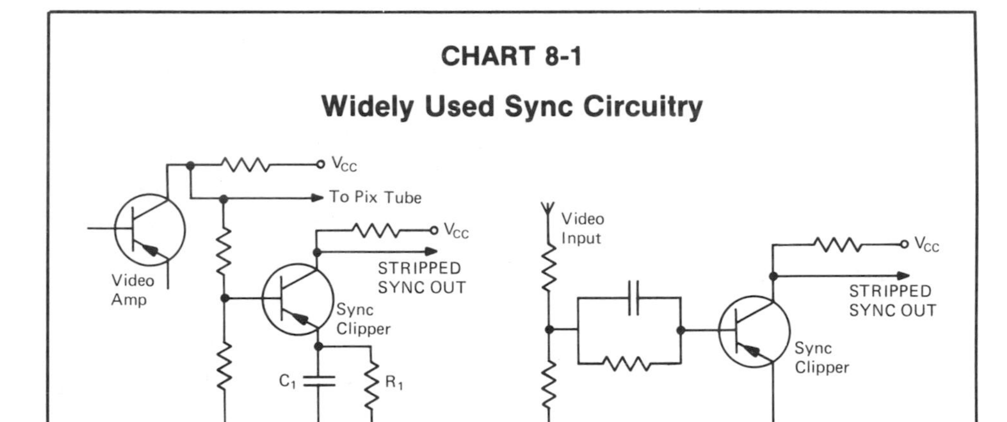

In this arrangement, signal-developed bias is produced in the emitter branch of the sync-clipper circuit.

In this arrangement, signal-developed bias is produced in the base branch of the sync-clipper circuit.

The sync section commonly starts from the video amplifier which is direct-coupled to the sync-clipper transistor. Although the complete video signal is applied to the sync clipper, only the sync tips normally appear in the stripped-sync output.

To maintain correct clipping action over a wide range of signal-input levels, signal-developed bias is employed in these typical arrangements. When a negative-going video signal is applied to the sync clipper transistor, the base-emitter junction becomes

CHART 8-1 CONTINUED

forward-biased and the transistor conducts. In turn, C_1 charges toward the peak of the applied voltage.

Note that C_1 charges rapidly, and at the end of the sync pulse the base is biased less negative. C_1 then starts to discharge slowly through R_1.

In normal operation, C_1 loses only a fraction of its charge between sync pulses. Or, the sync-clipper transistor normally remains cut off until the next sync pulse arrives and replenishes the charge on C_1.

Note that each time a sync tip is applied to the sync-separator transistor, collector current flows and an amplified sync tip appears in the stripped-sync output.

In normal operation, the reverse bias voltage produced by the signal-developed bias action is a fixed percentage of the peak input signal voltage. This is just another way of saying that in normal operation, the bias voltage is automatically adjusted for changes of input signal voltage level.

Note that although the bias voltage is automatically adjusted for changes of input signal level, subsequent limiting is necessary for proper system operation because the amplitude of the stripped sync output pulse depends somewhat upon the input signal amplitude. For example, if the input signal voltage decreases by 1 volt, and the voltage on C_1 drops to 80 percent of this peak voltage between pulses, the forward bias applied to the emitter during the next sync pulse is $1 - 0.8$ volt, or 0.2 volt. Then, if the input signal is doubled, the charge remaining on C_1 is 1.6 volts and the forward bias is 0.4 volt.

CHART 8-2

TV Troubleshooting with the Oscilloscope

Signal-channel checks with the oscilloscope usually start at the picture-detector output. Trouble symptoms are as follows:

1. The composite video signal may be weak—in other words, its peak-to-peak voltage may be less than 1 volt.
2. The composite video signal may be distorted—for example, the sync tips might be compressed or clipped. The white region of the camera signal might be compressed or clipped.

CHART 8-2 CONTINUED

3. Sometimes hum interference is present in the video signal waveform.
4. If the tips of the vertical sync pulses are not level with the tips of the horizontal sync pulses, frequency distortion is present.
5. A double-image video waveform indicates the presence of a ghost signal.

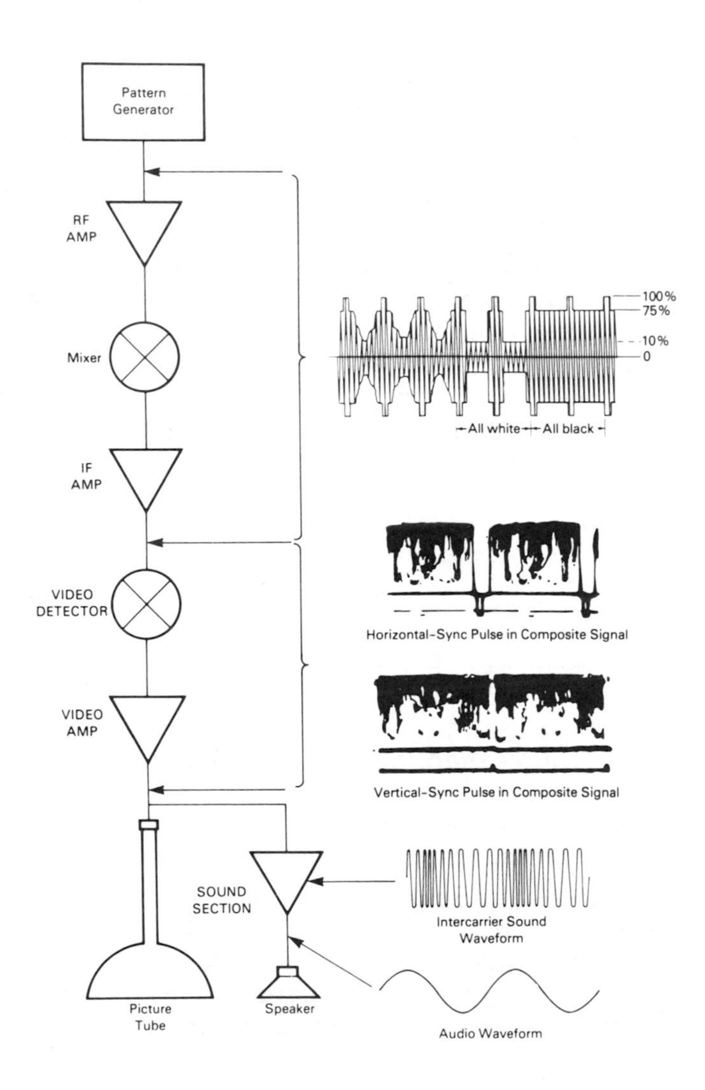

CHART 8-2 CONTINUED

Positive-going Sync Pulses

Negative-going Sync Pulses

The polarity of the video signal depends on the polarization of the video detector diode. (If a replacement diode is accidentally reversed in polarity, the video-signal polarity will be reversed.)

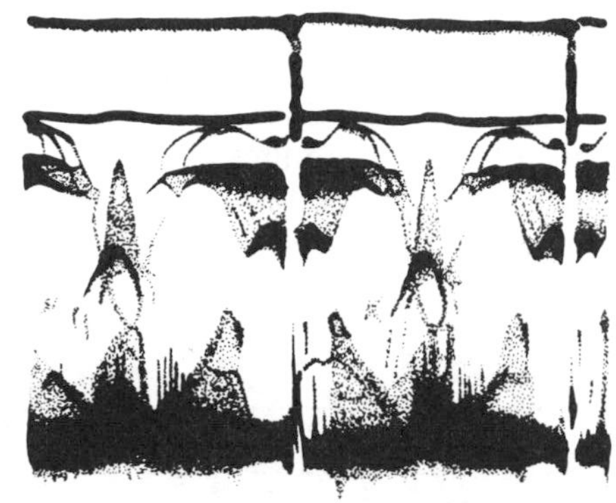

This is a normal composite video waveform, displayed on 30-Hz horizontal deflection. The sync is positive-going. There is no evidence of low-frequency attenuation or high-frequency attenuation, inasmuch as the tips of the vertical-sync pulse are level with the tips of the horizontal sync pulses.

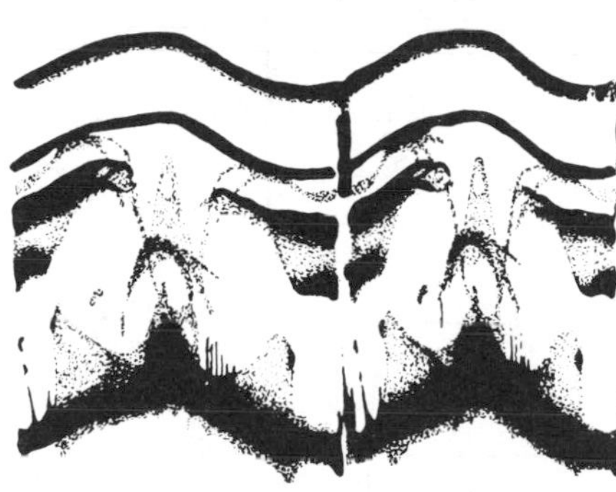

This is a composite video waveform with 60-Hz hum interference. The hum voltage usually results from a defective power-supply filter.

Hum interference sometimes has a 120-Hz frequency. Twice as many hum cycles appear in the pattern, in this case.

(The pattern will writhe if the video signal has a network program as its source.)

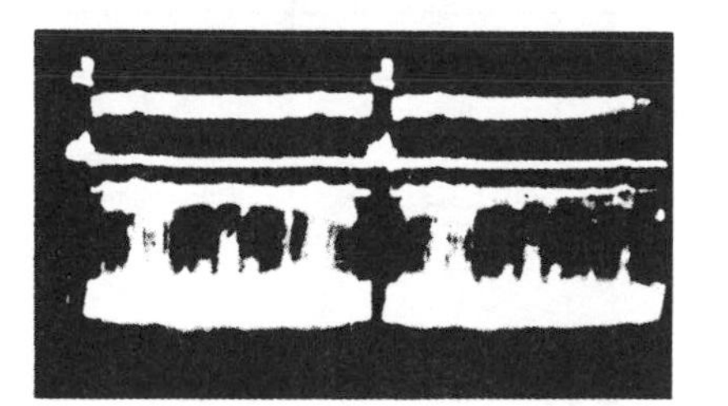

"Sync pushing" is a form of frequency distortion in which the tips of the vertical-sync pulse are higher than the tips of the horizontal sync pulses. The distortion results from improper alignment of the IF tuned circuits wherein the low video frequencies are peaked up excessively.

CHART 8-2 CONTINUED

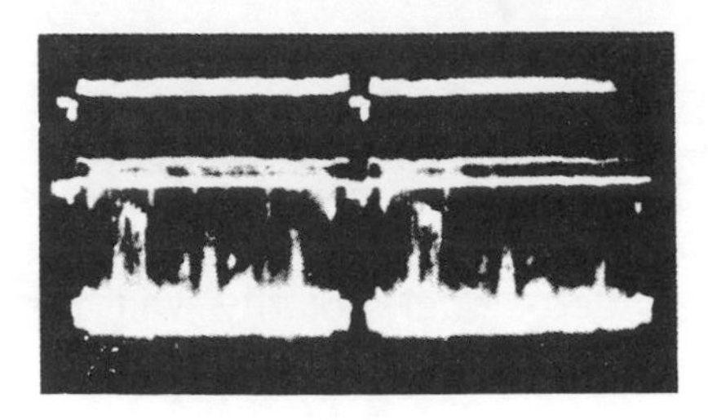

"Sync punching" is a form of frequency distortion in which the tips of the vertical sync pulse are lower than the tips of the horizontal sync pulses. It results from improper IF alignment. Sync punching can upset sync-clipper action and impair sync lock.

Troubleshooting Hint: A trouble symptom of poor picture reproduction does not necessarily point to trouble in the video signal channel. The cause of the trouble is frequently located in the filter section of the power supply. Deteriorated filter capacitors can result in poor picture quality, "touchy" sync, and unstable receiver operation.

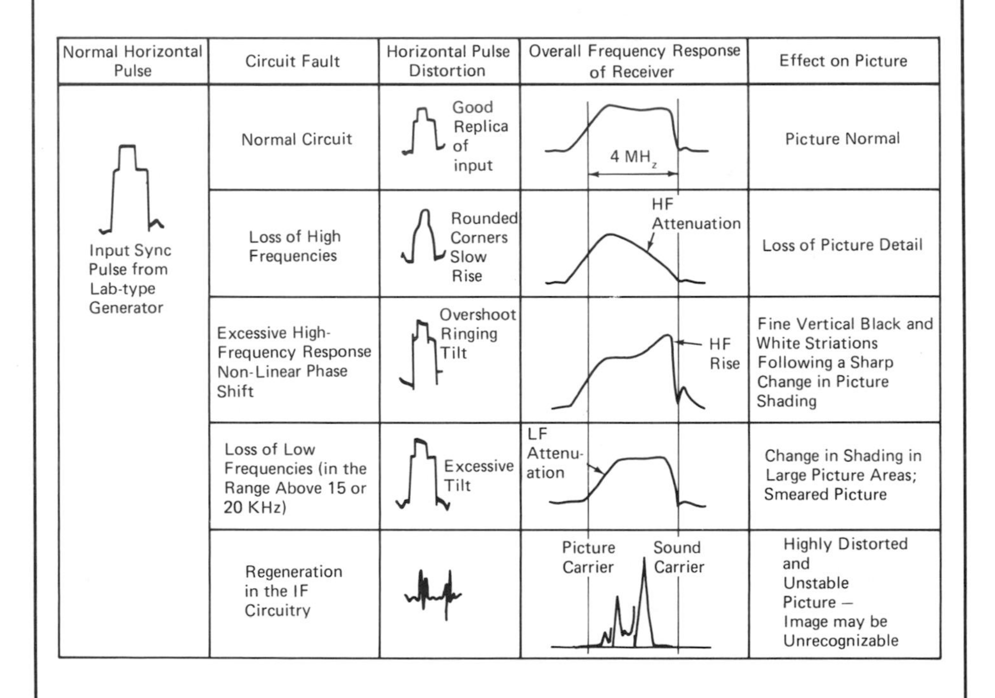

Normal Horizontal Pulse	Circuit Fault	Horizontal Pulse Distortion	Overall Frequency Response of Receiver	Effect on Picture
Input Sync Pulse from Lab-type Generator	Normal Circuit	Good Replica of input	4 MHz	Picture Normal
	Loss of High Frequencies	Rounded Corners Slow Rise	HF Attenuation	Loss of Picture Detail
	Excessive High-Frequency Response Non-Linear Phase Shift	Overshoot Ringing Tilt	HF Rise	Fine Vertical Black and White Striations Following a Sharp Change in Picture Shading
	Loss of Low Frequencies (in the Range Above 15 or 20 KHz)	Excessive Tilt	LF Attenu-ation	Change in Shading in Large Picture Areas; Smeared Picture
	Regeneration in the IF Circuitry		Picture Carrier Sound Carrier	Highly Distorted and Unstable Picture — Image may be Unrecognizable

Waveforms reproduced by special permission of Reston Publishing Co. and Walter Folger, from Radio, TV, and Sound System Diagnosis and Repair.

USING DUAL-TRACE OSCILLOSCOPES

Dual-trace oscilloscopes facilitate TV troubleshooting procedures inasmuch as two waveforms can be displayed simultaneously. For example, the composite video signal can be displayed with the sync-separator output signal. In turn, the troubleshooter can observe the effect on both waveforms of AGC adjustments, of signal input level variation, and of sync-clipper output with variation in camera-signal background brightness.

WAVEFORM TOLERANCES

There is a reasonable tolerance on TV waveforms; for example, an amplitude tolerance of ±20 percent on rated value is usually permissible. Tolerances on waveshape (distortion) are more involved, and usually entail the troubleshooter's practical experience.

IMPEDANCE QUICK CHECKS IN SYNC CIRCUITRY

When you are troubleshooting a dead receiver, it is helpful to measure resistances to ground with a low-power ohmmeter. Resistance measurements assist in pinpointing defective transistors with excessive junction leakage, and capacitors with excessive leakage. However, another type of quick check is required to localize open capacitors—if a sync circuit passes a resistance checkout, it is good practice to follow up with impedance quick checks.

Impedance quick checks, such as those described in Chart 8-3, are very informative when made on a comparative basis with respect to a similar receiver that is in normal operating condition. Note that the test data are subject to reasonable tolerances, such as ±10 percent. A ±20 percent difference in readings can be regarded with suspicion; a ±40 percent difference is a sure sign of circuit trouble.

CHART 8-3

Impedance Quick Check in Sync Circuitry

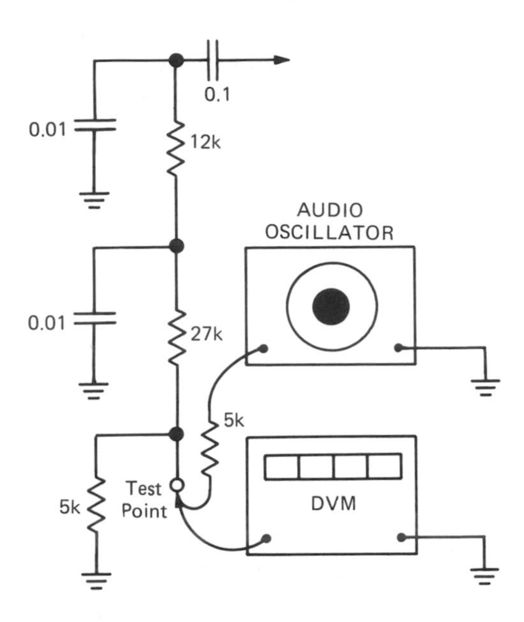

Sync circuitry contains RC configurations that have particular time constants. In turn, impedance values are as important as resistance values. In-circuit measurements of impedance values are most helpful when made with respect to a similar receiver on a comparison basis.

In this example of sync-separator circuitry, a test point is provided for DC-voltage measurement. However, in the case of a dead receiver, troubleshooting involves resistance and impedance measurements.

The impedance quick check is made by injecting a test signal from an audio oscillator through a 5-kilohm resistor and checking the AC voltage at the injection point.

This quick check is most informative when made on a comparison basis. Observe that the 0.01μF capacitors in the vicinity of the test point are returned to ground. Accordingly, this is a frequency-responsive circuit.

Practical Example: The audio oscillator was set to 500 Hz with an output level that provided a 40-mV reading on the DVM. Then, the

CHART 8-3 CONTINUED

test frequency was increased to 10 kHz, and the DVM reading dropped to 4 mV. This reduction in voltage at the test point was the result of the AC shunting action of the fixed capacitors.

When the 0.01μF capacitor nearest the test point was open, the DVM reading was 6 mV at 10 kHz. This was an increase of 50 percent in AC voltage, and clearly indicated a trouble condition.

Note that when the 0.01μF capacitor farthest from the test point was open, the DVM reading was only slightly greater than when the capacitor was normal. In other words, this capacitor is separated from the test point by so much resistance that a useful test cannot be made. This is an example of the principle that impedance quick checks should be made at more than one test point in a suspected circuit.

In most situations, it is sufficient to make impedance quick checks at the base, emitter, and collector terminals of the sync-section transistors.

9

Progressive Television Troubleshooting Methods

*Sync Separator with Automatic Time Constant * Troubleshooting a Double T-C Sync Separator Stage * Noise-Switched Sync Separator * Intermittent Monitoring * Quick-Check Capacitance Indicator and Comparator * Constant-Current Impedance Quick Checker * Troubleshooting Vertical Deflection Circuitry * Yoke Coupling * Test Instrument Application * Using Ohmmeter as Intermittent Signal Monitor*

SYNC SEPARATOR WITH AUTOMATIC TIME CONSTANT

"You can't know too much about circuit action when troubleshooting without service data" is an axiom in progressive methodology. One of the frequently encountered sync-separator circuits provides automatic double time-constant operation, as depicted in Figure 9-1.

This is a diode switching arrangement in which diode D automatically switches the long time-constant branch circuit in or out of the emitter return circuit as required by the presence or absence of noise pulses. During the time that the sync tip is inputted, the transistor conducts, and a negative pulse voltage appears at the emitter terminal. This voltage places a forward bias on diode D, and diode conduction switches the long time-constant circuit into the emitter leg.

This long time constant provides an optimum separation level for the prevailing signal level, and contributes to bias stability. Following passage of the sync tip, or noise pulse, the emitter voltage drops and the capacitor in the long time-constant branch starts to

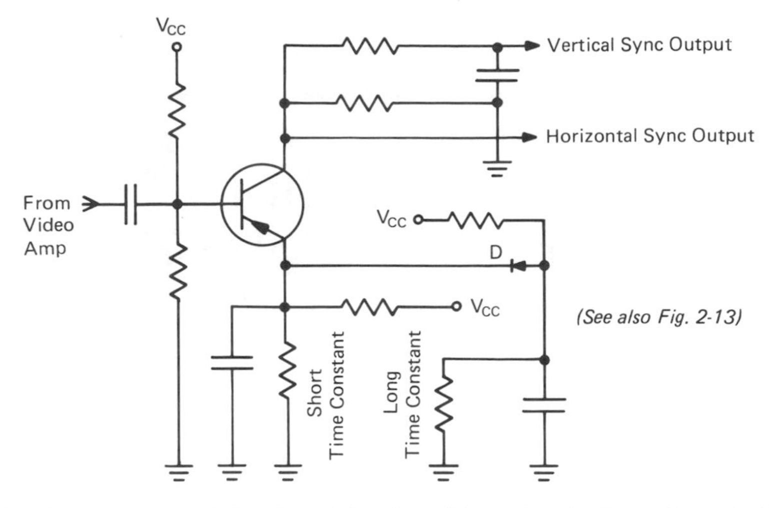

The time constant of the signal-developed bias circuit plays a key role in noise rejection by the sync separator and resulting freedom from picture tearing. Horizontal deflection circuits are best served by a sync separator with a short time constant, which provides rapid recovery of normal bias after a large peak-value noise pulse has passed. On the other hand, vertical deflection circuits are best served by a long time constant, wherein a large noise pulse does not cause a deep dip in the output signal that could falsely trigger the vertical oscillator. Although two sync-separation circuits can be employed, this is not an economical design. Most receivers utilize either a compromise-design sync separator, or a double time-constant configuration, as exemplified above.

Figure 9-1 A widely used double time-constant sync-separator configuration.

discharge. However, the capacitor voltage remains more negative than the emitter voltage and diode D is reverse-biased.

The long time-constant branch is effectively disconnected from the emitter leg at the end of the charging pulse; it cannot back off or block the sync separator. Note that the short time-constant branch (which was charged in parallel with the long time-constant branch) discharges rapidly and the sync separator recovers quickly.

It is evident that the long time-constant branch normally functions to absorb the energy of large noise pulses without causing significant back-off in the separator transistor.

TROUBLESHOOTING A DOUBLE T-C SYNC SEPARATOR STAGE

When troubleshooting without service data, the technician may or may not have a similar receiver in normal working condition. Also, the sync section in the bad receiver may have more or less response, or it may be completely dead. These considerations must be taken into account in preliminary troubleshooting procedures.

In any case, an in-circuit transistor tester may be used to determine whether the transistor is workable. If a comparison receiver is at hand, and the bad receiver is not dead, DC-voltage measurements may be made. On the other hand, if the bad receiver is dead, only comparison resistance measurements can be made at the outset.

If the bad receiver passes the in-circuit resistance checks, it is good practice to follow up with impedance quick checks of the separator stage, as explained in Chapter 8. Open capacitors, for example, show up on an impedance quick check.

In the event that the sync section in the bad receiver has more or less response, and a comparison receiver is available, oscilloscope waveform tests can be very informative. Of course, an off-value and/or distorted waveform indicates that something is wrong in the associated circuit; the technician should endeavor to analyze the incorrect waveform in order to quickly close in on the fault.

It follows from the basic examples of waveform analysis shown in Chapter 8 that this is a comparatively sophisticated technique, and one that requires considerable practical experience. When a comparison receiver is not at hand, the troubleshooter is thrown upon his or her own resources. You can't know too much about circuit action here. (See also Chart 9-1.)

NOISE-SWITCHED SYNC SEPARATOR

Another frequently encountered sync-separator arrangement is the noise-switched configuration exemplified in Figure 9-2. This is a widely used arrangement. Like the double time-constant circuit, it provides considerable immunity to noise pulses and resulting picture

CHART 9-1

Quick-Check Capacitance Indicator and Comparator

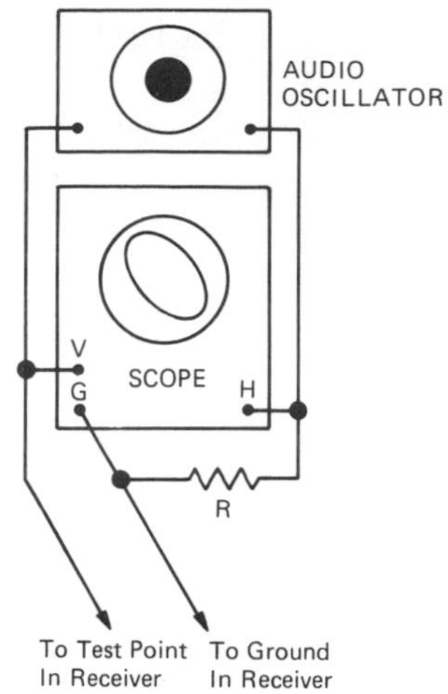

(As a comparator, this checker shows whether a capacitor has lost much of its capacitance)

Open or shorted capacitors can be easily spotted with this test arrangement.

An AC test voltage is applied at a chosen test point in the receiver circuitry. This applied voltage is fed to the vertical-input channel of the scope. The current drawn by the circuitry under test produces a corresponding voltage drop across resistor R, which is fed to the horizontal-input channel of the scope.

The resulting voltage/current Lissajous pattern on the scope screen shows whether the receiver circuitry under test is resistive or capacitive and its relative proportions of resistance and capacitance:

1. When a resistor is connected between the two test leads, a diagonal line is displayed on the scope screen.
2. When a capacitor is connected between the two test leads, an ellipse (or circle) is displayed on the scope screen.
3. If a shorted capacitor is connected between the two test leads, a horizontal line is displayed on the scope screen.
4. If an open capacitor is connected between the two test leads, a vertical line is displayed on the scope screen.
5. If a capacitor shunted by resistance is connected between the two test leads, a diagonal ellipse is displayed on the scope screen.
6. If a capacitor in series with resistance is connected between the two test leads, a diagonal ellipse is displayed on the scope screen.
7. When a capacitor shunted by resistance and in series with resistance is connected between the two test leads, a diagonal ellipse is displayed on the scope screen.

CHART 9-1 CONTINUED

Note that the audio oscillator must have balanced (push-pull or double-ended) output.

Resistor R can have any value that provides adequate horizontal deflection.

A high-sensitivity scope is used, with low-level output from the audio oscillator. (This ensures that transistors in the circuit under test will not be turned on.)

The frequency of test is not critical, although it should not be so low or so high that the capacitance in the circuit under test produces an excessively narrow ellipse.

Note that if the test leads are applied in an inductive circuit an ellipse is displayed—this ellipse tilts "up-hill" instead of "down-hill."

This quick checker indicates whether the circuit under test is resistive or capacitive. When used for comparative tests in a similar receiver, the quick checker also shows whether the resistive and capacitive components are *the same,* or *not the same.*

tearing in normal operation. When a picture-tearing trouble symptom occurs, the noise-rejection circuit should be checked first.

An in-circuit transistor tester can be used to determine whether transistors are workable. If a similar good receiver is available, troubleshooting is considerably facilitated. Impedance quick checks, DC-voltage measurements, resistance measurements, and waveform checks can be made to close in on the fault. If a comparison receiver is not at hand, the troubleshooter must proceed by dead reckoning, as previously noted.

INTERMITTENT MONITORING

This intermittent tough dog problem presents aggravated difficulties in troubleshooting procedures, and an audio-tone monitor is often helpful. An intermittent condition does not always react to tapping, heating, cooling, or switching tests. In such a case, the troubleshooter can only let the receiver "cook," and wait for the intermittent to show up. This can sometimes be a long wait.

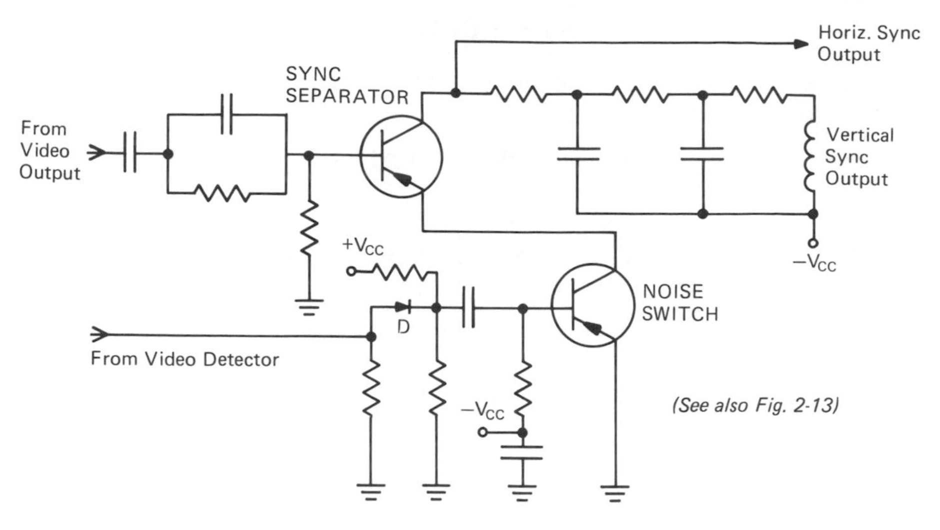

This type of noise-rejecting sync separator is also called a "hole puncher" because the sync-pulse train is temporarily cut off following arrival of a large noise pulse. This circuit action is preferable to admission of the noise pulse, because the horizontal-oscillator section is more tolerant of missing sync pulses than of large noise pulses. Diode D is called a noise separator; it becomes forward-biased when a large noise pulse appears, and in turn the noise-switch transistor is cut off, thereby effectively open-circuiting the emitter lead of the sync-separator transistor, which cuts off the sync separator momentarily. In turn, the large noise pulse is "punched out," along with one or two horizontal sync pulses. Because of the "flywheel effect" of the following horizontal-AFC section, the picture does not tear. Note that the noise separator is driven by the video detector, whereas the sync separator is driven by the video amplifier. Accordingly, a noise pulse will arrive at the noise-switch transistor slightly before it arrives at the sync-separator transistor. This propagation delay provides adequate time for proper noise-switch operation.

Figure 9-2 Sync-separator stage with a noise-switch control.

The intermittent monitor depicted in Figure 9-3 may be connected into a circuit that is suspected of being intermittent, and in turn will alert the troubleshooter to any voltage change by means of a change in output tone. The monitor is responsive to either a DC or an AC voltage change, or both.

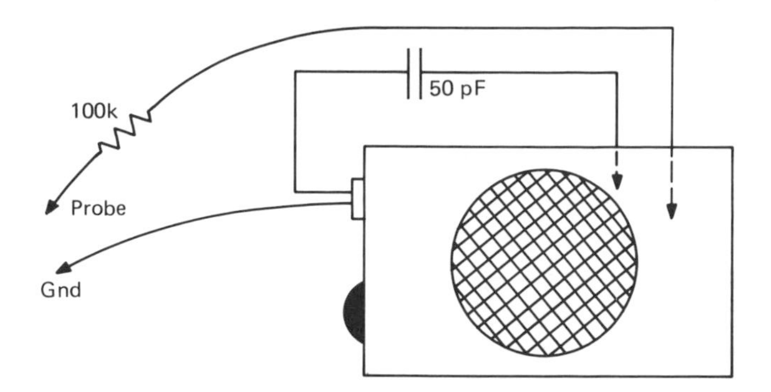

Technical Note: This tone-output intermittent monitor utilizes a miniamp/speaker unit such as the Archer (Radio Shack) 277-1008A. It is operated as a tone generator by connecting a 50-pF capacitor from the blue-lead terminal of the mini speaker to the input terminal of the amplifier. When the volume control is advanced, a low tone output is generated.

To operate the tone generator as an intermittent monitor, a 100-kilohm resistor is used as a probe. The resistor is connected to a test lead, which in turn connects to pin 8 on the integrated circuit inside of the amplifier.

In application, the probe and ground leads are connected into the circuit that is to be monitored. The monitor outputs a tone that remains constant in pitch as long as the voltage at the test point remains constant. On the other hand, any voltage change results in a change in pitch of the tone from the monitor.

Figure 9-3 A simple tone-output monitor for intermittent troubleshooting.

CONSTANT-CURRENT IMPEDANCE QUICK CHECKER

The constant-current impedance quick checker depicted in Figure 9-4 is quite informative, because it can be calibrated to indicate impedance values, and thereby measure impedances at audio frequencies. Its principle of operation is to provide a virtually constant value of AC test current, regardless of the impedance under test. (Recall that $R = E/I$, $X = E/I$, and $Z = E/I$. In turn, if I is equal to unity, then a resistance, reactance, or impedance value is equal to the voltage drop across the component under test.)

As an illustration of resistance, reactance, and impedance measurements, the test leads may be connected across a 1-kilohm

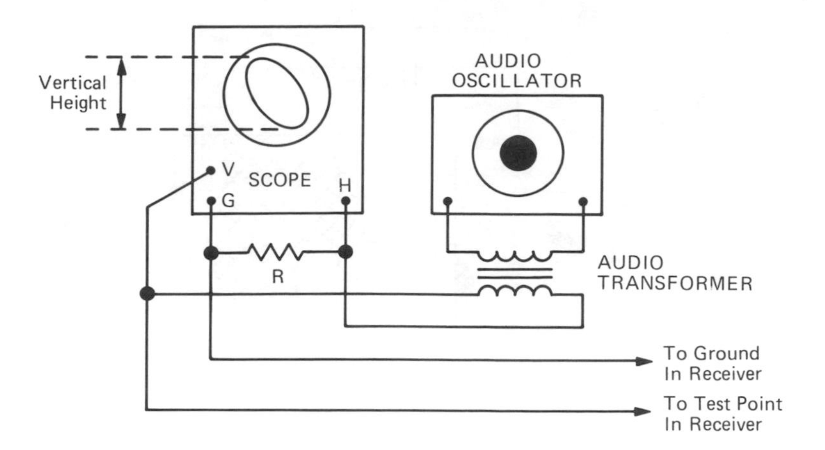

An audio transformer is used in this arrangement to step up the output voltage from the audio oscillator. The resistance of R is comparatively high; you may use a potentiometer in order to adjust the horizontal deflection, as shown below:

To H input
To G — To transformer

Note that perfect circles and true ellipses will not be displayed on the scope screen if there are harmonics in the audio-oscillator output, or if the audio transformer distorts the sine waveform.

(Although distortion is undesirable, it does not invalidate the reliability of comparison tests.)

This quick checker uses a comparatively high source voltage and a comparatively high source resistance R. In turn, a constant-current impedance check is provided. The advantage of this arrangement is that the vertical height of the screen pattern indicates the impedance of the circuit under test. (The scope can be calibrated to indicate impedance values in ohms.) Note that impedance is related to frequency—quick checks may be made at a test frequency of 100 Hz, 1 kHz, or 10 kHz. For example, the reactance (impedance) of a 0.01μF capacitor is 1600 ohms at 10 kHz, 16 kilohms at 1 kHz, and 160 kilohms at 100 Hz. The test frequency is chosen to provide a suitable pattern on the scope screen. Note that the horizontal deflection remains practically constant, regardless of the test frequency, or the impedance across which the test leads are connected.

Note that the value of R should be high enough (and the output from the audio oscillator should be low enough) that you do not apply more than 500 mV across the circuit under test. This ensures that semiconductor junctions will not be turned on.

Figure 9-4 Constant-current quick check impedance indicator and comparator.

resistor, and the scope may be adjusted for 1 inch of vertical deflection. Then, if you connect a 2-kilohm resistor between the test leads, you will observe 2 inches of vertical deflection. Or, if you connect a 500-ohm resistor between the test leads, you will observe one-half inch of vertical deflection.

Next, if you connect a 0.01μF capacitor between the test leads and set the audio oscillator to 10 kHz, you will observe 1.6 inches of vertical deflection of the scope screen. Again, if you set the audio oscillator to 1 kHz and reduce the vertical attenuator to 0.1 of its previous setting, you will again observe 1.6 inches of deflection on the scope screen.

Note that you obtained 1.6 inches of vertical deflection in the foregoing tests because the reactance of the capacitor was 1600 ohms at 10 kHz, and 16 kilohms at 1 kHz. (Note also that the horizontal deflection was the same in all of the foregoing tests—because this is essentially a constant-current test method.)

Next, if you connect a 0.01μF capacitor in series with a 1600-ohm resistor between the test leads, you will obtain 2.26 inches of vertical deflection, because the impedance of the series RC combination is 2260 ohms, approximately.

Then if you connect a 0.05μF capacitor in parallel with a 1600-ohm resistor between the test leads, you will obtain 1.42 inches of vertical deflection, because the impedance of the parallel RC combination is 1420 ohms, approximately.

To briefly recap the screen patterns obtained in various tests, note that:

1. When a resistor is connected between the two test leads, a diagonal line is displayed. (The "slope" of the line is disregarded—the troubleshooter observes the amount of vertical deflection to determine the resistance value.)
2. When a capacitor is connected between the two test leads, an ellipse (or circle) is displayed. (The amount of horizontal deflection is the same as before and is disregarded—the troubleshooter observes the amount of vertical deflection to determine the reactance value.)
3. If a shorted capacitor is connected between the two test leads, a horizontal line is displayed—the troubleshooter recognizes that the lack of any vertical deflection indicates a short-circuit.

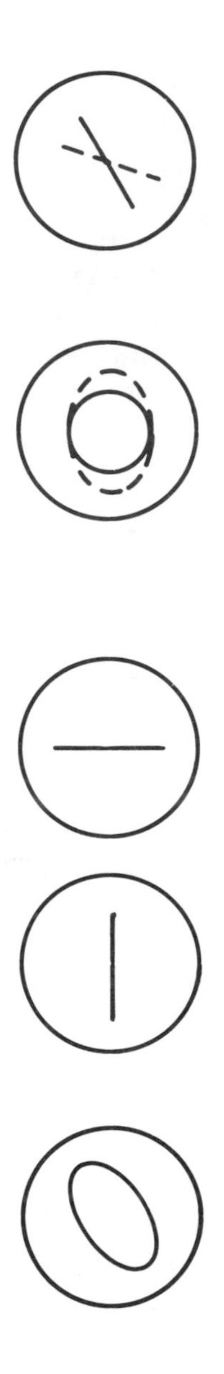

A diagonal line indicates that resistance is connected between the test leads. The vertical deflection in the display indicates the number of ohms of resistance. (The inclination of the line depends on the value of source resistance you have utilized.)

A circle or an ellipse with a vertical major axis indicates that capacitance is connected between the test leads. The vertical deflection in the display indicates the number of ohms of reactance. (Whether a circle or an ellipse happens to be displayed depends on the value of source resistance you have utilized.)

A horizontal line indicates that there is a short-circuit between the two test leads. (The length of the horizontal line depends on the value of source resistance you have utilized.)

A vertical line indicates that the test leads are open-circuited. (Residual or stray capacitance between the test leads may cause a very narrow vertical ellipse to be displayed.)

A diagonal ellipse indicates that the test leads are connected across an impedance. The impedance might be an RC series combination, an RC parallel combination, or an RC series-parallel combination. (The resistive and reactive components are a function of the voltage/current phase angle, or the inclination of the ellipse. These components are ordinarily disregarded in practical troubleshooting procedures.)

Figure 9-5 Patterns displayed by impedance checker, showing Ohm's Law for AC.

4. If an open capacitor is connected between the two test leads, a vertical line is displayed—the troubleshooter recognizes that the lack of any horizontal deflection indicates an open circuit.
5. When a capacitor shunted by resistance is connected between the two test leads, a diagonal ellipse is displayed. (The amount of horizontal deflection is the same as before—the troubleshooter observes only the amount of vertical deflection to determine the impedance value.)
6. When a capacitor in series with resistance is connected between the two test leads, a diagonal ellipse is displayed. (The amount of horizontal deflection is the same as before—the troubleshooter observes only the amount of vertical deflection to determine the impedance value.)
7. If a capacitor shunted by resistance in series with resistance is connected between the two test leads, a diagonal ellipse is displayed. (The amount of horizontal deflection is the same as before—the troubleshooter observes only the amount of vertical deflection to determine the impedance value.)

Note in passing that the inclination (the amount that an ellipse "leans") indicates the phase angle between the voltage and current in the circuit under test. However, in practical troubleshooting procedures this phase angle is not measured, and is disregarded. (See Figure 9-5.)

TROUBLESHOOTING VERTICAL DEFLECTION CIRCUITRY

A troubleshooter needs to be familiar with the standard types of vertical deflection circuits and with their circuit action when troubleshooting without service data. A skeleton vertical-output configuration is shown in Figure 9-6. The vertical-deflection coils normally conduct a sawtooth current; a 17-inch picture tube requires a typical sawtooth current of 400 mA p-p.

As noted in Figure 9-6, the vertical deflection coils place an essentially resistive load on the collector of the output transistor. However, this is not a purely resistive load—it has an inductive component which cannot be neglected. Accordingly, a trapezoidal voltage waveform is required to drive a sawtooth current through the deflection coils. Briefly, this requirement is as follows:

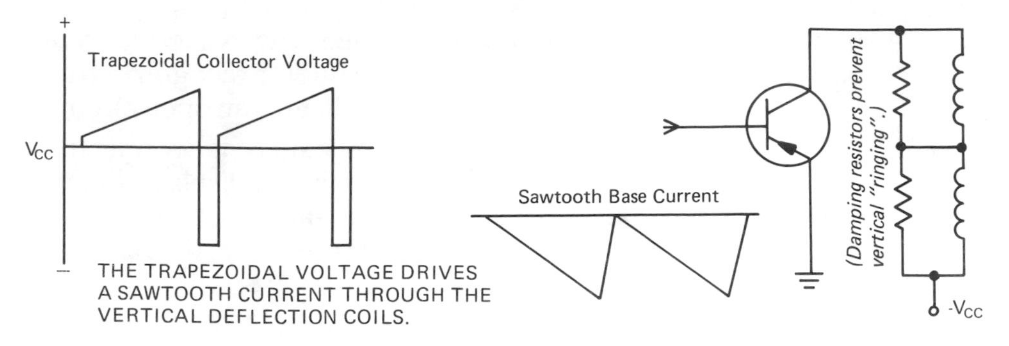

The vertical-output transistor is directly coupled to the vertical deflection coils in this widely used arrangement. The deflection coils place an effectively resistive load on the collector, and the circuit may be regarded as a class-A amplifier. During the retrace interval, the rapid current change produces a pulse of CEMF (counter electromotive force) across the deflection coils; this is called the retrace pulse. Since the sawtooth base waveform is negative-going, this retrace pulse does not cause the collector junction to become forward-biased. As depicted in the diagram, the voltage across the coils reverses its polarity during the retrace interval.

Note that the two damping resistors do not necessarily have the same resistive value, and that a thermistor may be connected in series with the two deflection coils. The thermistor compensates for the "hot" resistance of the coils.

Figure 9-6 Skeleton vertical-output configuration. Common-emitter operation provides high-power gain.

1. A sawtooth voltage will drive a sawtooth current through a purely resistive load.
2. A pulse (rectangular) voltage will drive a sawtooth current through a purely inductive load.
3. In turn, a combination of these two waveforms (a trapezoidal waveform) will drive a sawtooth current through a resistive load with a residual inductive component.

Yoke Coupling

The troubleshooter should also recognize the different types of coupling that are used between the vertical-output transistor and the vertical-deflection coils in the yoke. As explained subsequently, certain trouble symptoms will point to a fault in the coupling circuit. Technically, the skeleton configuration depicted in Figure 9-6 is oversimplified and impractical. A suitable coupling arrangement is necessary.

Note in Figure 9-6 that the sawtooth deflection current will vary from zero to a maximum value during the trace interval, and this deflection current does not reverse its direction of flow at any time over the complete current cycle. This is just another way of saying that all of the vertical deflection is confined to one-half the screen in a vertical direction.

Moreover, there is a large DC component accompanying the sawtooth current in the skeleton arrangement. This DC component serves no useful purpose and merely wastes power. Although vertical-centering means could be used to scan the entire screen vertically, the objection of wasted power remains. Therefore, *we will find two chief types of coupling arrangements for eliminating DC current flow through the vertical deflection coils:*

1. A vertical-output transformer may be found between the vertical-output transistor and the yoke, as shown in Figure 9-7(a).
2. A large series coupling capacitor may be found between the vertical-output transistor and the yoke, as shown in Figure 9-7(b). (Either arrangement is fairly easy to spot and provides a useful landmark in finding your way around the vertical-deflection section.)

TEST INSTRUMENT APPLICATION

Troubleshooting vertical deflection circuits involves DC voltage measurements and resistance measurements with a low-power ohmmeter. An impedance checker is very helpful, particularly when a similar receiver is available for comparison tests. Note that an impedance checker is applied in inductive circuits just as in capacitive circuits. The only difference is that elliptical patterns will "lean" to the right instead of to the left, when the checker is applied in an inductive circuit.

An impedance checker can be very helpful when a replacement yoke must be picked out from random stock. In other words, the DC

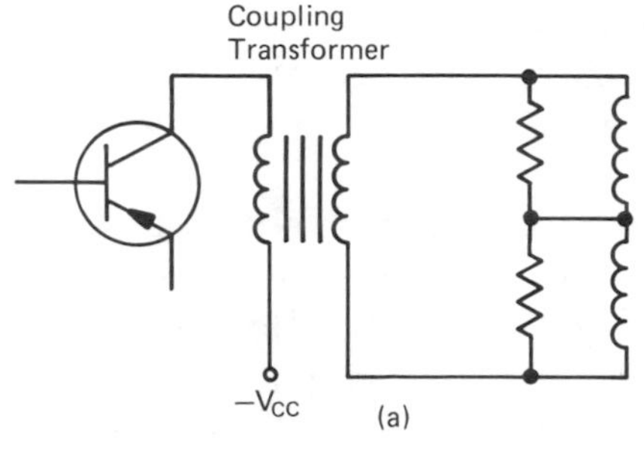

(a) The primary winding of the vertical-output transformer has comparatively low resistance, thereby minimizing the DC power loss in the vertical-output circuit.

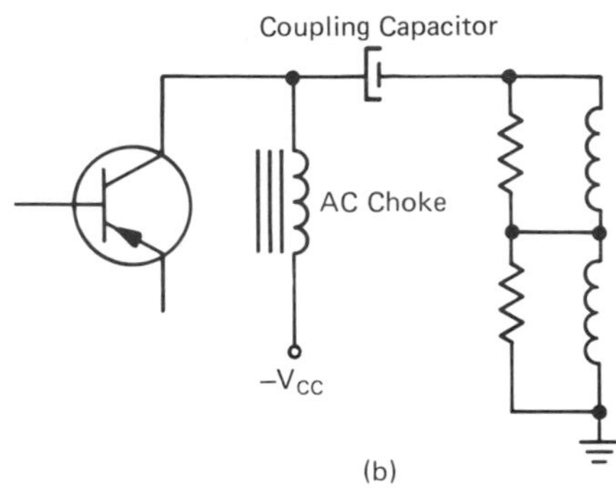

(b) The AC choke has comparatively low resistance in order to minimize the DC power loss in the vertical-output circuit. A large coupling capacitor is used, such as 1000μF.

Another output arrangement that may be found consists of the circuit shown in (b), with the coupling capacitor omitted. Since the AC choke has comparatively low resistance, most of the DC component flows through the choke. However, some of the DC component flows through the deflection coils. In turn, we will find centering magnets provided to compensate for the decentering action of this DC current through the coils.

Preliminary troubleshooting of the vertical-output section should start with a comparison temperature check of the vertical-output transistor. If it is running too hot, the total collector current is excessive; if it is running too cold, the total collector current is deficient.

Figure 9-7 Widely used coupling arrangements. (a) Vertical-output transformer; (b) Coupling capacitor and choke.

resistance of the deflection coils is not sufficient identification for suitable replacement. An impedance checker, when used on a comparison basis, can precisely identify a suitable replacement yoke.

An oscilloscope can often help to close in on a deflection-circuit fault when used for comparison waveform tests. However, a scope cannot be used in a dead deflection circuit. Note that a TV analyzer can be utilized to inject drive waveforms and deflection waveforms when signal-substitution tests are needed.

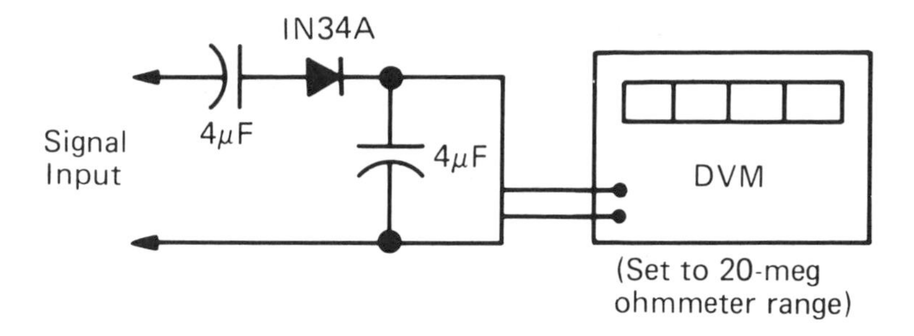

The signal-input leads are applied at a test point in the receiver. The DVM is operated on its highest resistance range, with the beeper turned on. If the signal stops, the beeper is activated.

This type of intermittent signal monitor employs the continuity indicator function of a DVM with a beeper. The DVM leads are connected with the positive (red) lead at the cathode terminal of the diode. The monitor operates as follows:

Note that when the DVM is first connected across the output capacitor, the ohmmeter starts to charge the capacitor, and the beeper sounds. After the capacitor charges up to the ohmmeter test voltage, the beeper becomes silent.

Next, when the signal-input leads are applied at a test point in the receiver, with signal present, the beeper remains silent. In other words, the output capacitor is charged to an even higher positive voltage by the rectified signal.

In the meantime, the input coupling capacitor is charged in the opposite polarity by the rectified signal.

If and when the signal stops, the output capacitor can no longer maintain its charge, because it proceeds to discharge through the reverse resistance of the germanium diode into the oppositely charged input coupling capacitor.

As a result, the beeper sounds as the ohmmeter proceeds to charge up the output capacitor.

The beeper continues to sound for some time, until the output capacitor is finally charged up to the ohmmeter test voltage.

Note that proper operation of the intermittent signal monitor requires capacitors that have very high leakage resistance (insulation resistance). GE pyranol capacitors are suitable.

To avoid possible damage to the ohmmeter from comparatively high signal voltages, it is advisable to use a DVM with protected resistance ranges.

Figure 9-8 Intermittent signal monitor utilizing ohmmeter beeper.

Using Ohmmeter as Intermittent Signal Monitor

The intermittent monitor described above operates on the basis of DC voltage interruption. However, some intermittent conditions involve signal stoppage with no interruption of DC voltage. In turn, the troubleshooter needs to use an intermittent signal monitor such as that shown in Figure 9-8.

This arrangement provides audible indication from the ohmmeter beeper if and when the signal stops. Thus, it liberates the troubleshooter while waiting for an intermittent to occur. The DVM may be operated to advantage on its 20-megohm range; if it is operated on a low resistance range, the beeper will sound for only a short time after the signal stops. On the other hand, it will sound for quite a while when the DVM is operated on its 20-megohm range. (See also Chart 9-2.)

CHART 9-2

What a Test Instrument "Sees" in a Typical Circuit

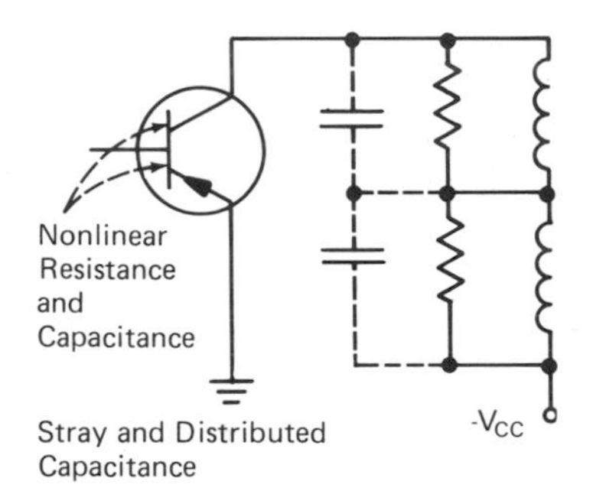

A vertical-output circuit is a network comprising inductance, stray, distributed, and junction capacitance, linear resistance, nonlinear resistance, and voltage (with associated current, power, and heat).

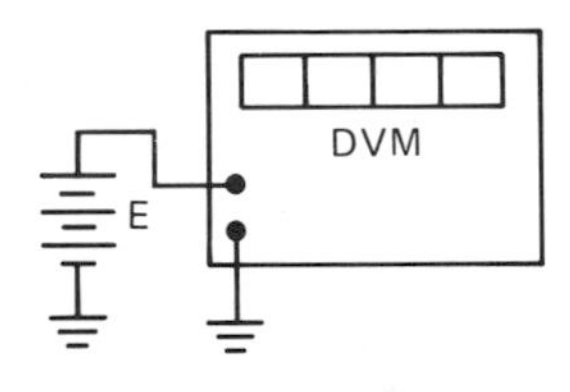

A DC voltmeter sees a voltage source and ignores the other electrical characteristics of the circuit.

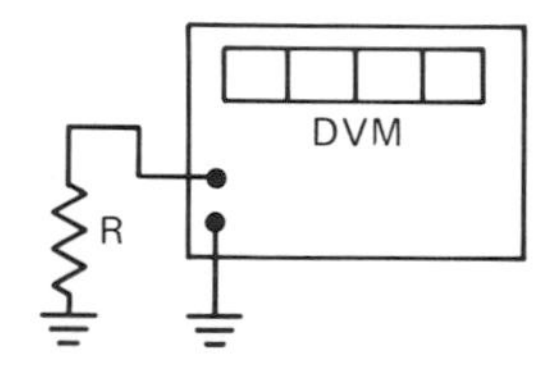

An ohmmeter sees a resistance and ignores the other electrical characteristics of the circuit.

CHART 9-2 CONTINUED

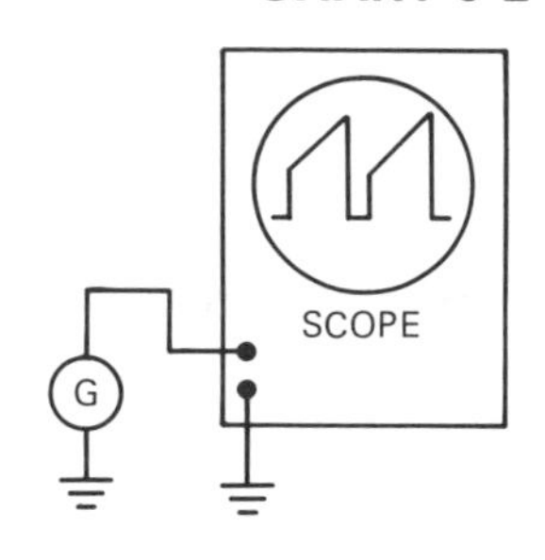

An oscilloscope sees a waveform generator and ignores the other electrical characteristics of the circuit.

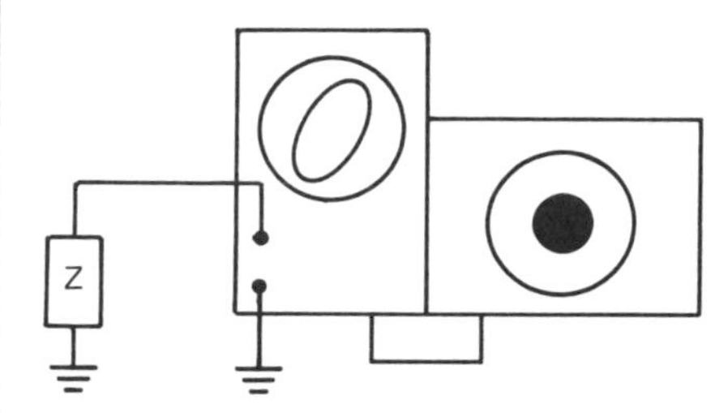

An impedance checker sees an impedance and ignores the other electrical characteristics of the circuit.

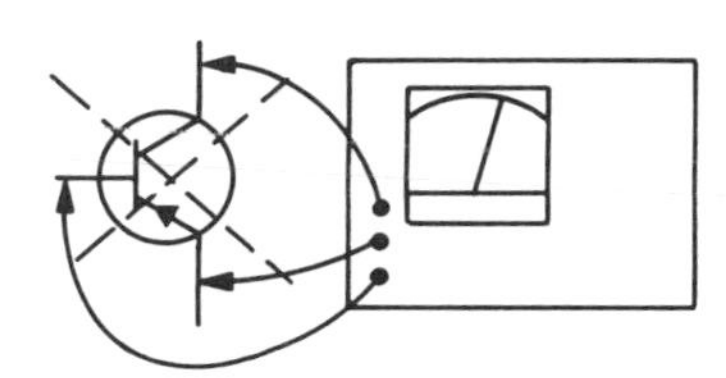

An in-circuit transistor tester sees a workable or an unworkable transistor and ignores the other electrical characteristics of the circuit.

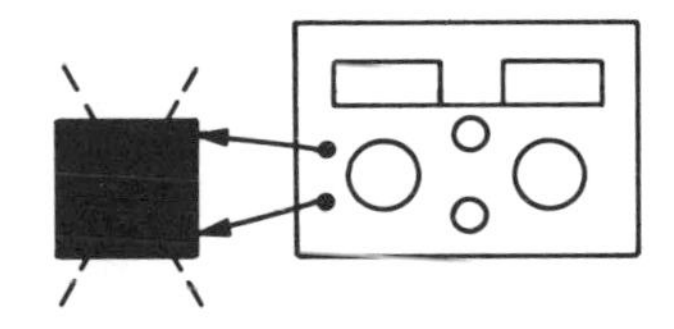

A TV analyzer sees a workable or an unworkable black box and ignores the other electrical characteristics of the circuit.

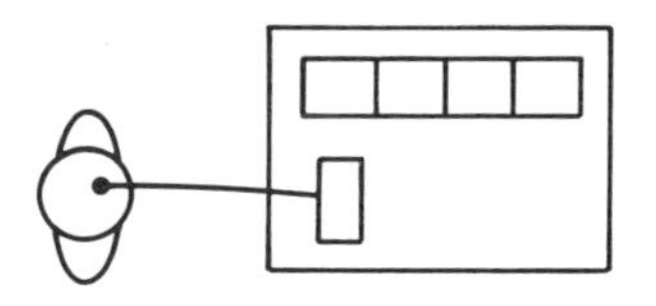

A temperature checker sees a heat source and ignores the other physical characteristics of the circuit.

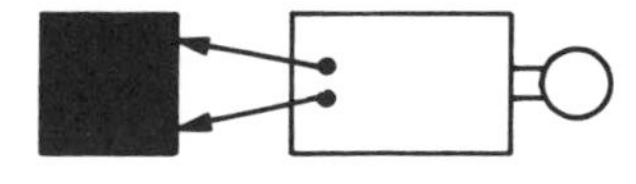

An intermittent monitor sees the presence or absence of a voltage in a black box.

Nanosiemens Units

DVMs that provide a 20-megohm resistance range may also have a 200 nanosiemens (nS) range for measuring resistances above 20 megohms. The nanosiemens range is a conductance range. Conductance is the reciprocal of resistance, and indicates permissiveness to electron flow. Thus, one siemens is the reciprocal of one ohm; 30 megohms is the reciprocal of 33.3 nS; 0.1 nS is the reciprocal of 10,000 megohms.

The chief advantage of the nanosiemens unit is its comparatively small numerical value, compared with the large numerical values of high resistances expressed in megohms. Thus, a nanosiemens is the reciprocal of 1000 megohms. Stated otherwise, a nanosiemens is the reciprocal of 10^{-9} ohms.

10

Follow-Up Television Troubleshooting Methods

*Vertical-Sweep Buffer Stages * Point-to-Point Impedance Checker * Common-Emitter Buffer Stage * Vertical-Oscillator Operation * Horizontal Blocking Oscillator * Horizontal AFC Stage * Beat Aural Signal Tracer * Horizontal Driver Stage * Horizontal-Output and High-Voltage Section * Comparison Temperature Checking*

VERTICAL-SWEEP BUFFER STAGES

It is helpful to recognize the basic types of vertical-deflection circuitry when troubleshooting without service data. A widely used vertical-sweep buffer arrangement is depicted in Figure 10-1. In normal operation it places only a light load on the sawtooth source, and provides substantial drive current to the vertical-output transistor.

Troubleshooting vertical-sweep buffer stages involves few difficulties when a similar receiver in normal working condition is available for comparison tests. DC voltage measurements and resistance measurements with a low-power ohmmeter will often turn up faults such as leaky or shorted capacitors, or defective transistors. (An in-circuit transistor tester can be used to follow up a suspicion of a bad transistor.)

When a stage is stone dead, DC voltage measurements may provide little or no useful data, and resistance measurements can be advantageously followed up with impedance checks. Impedance checks are usually quite informative when evaluated on a comparison basis. A point-to-point impedance checker such as that shown in Chart 10-1 permits the troubleshooter to check the impedance between two points in a circuit, both of which are above ground potential.

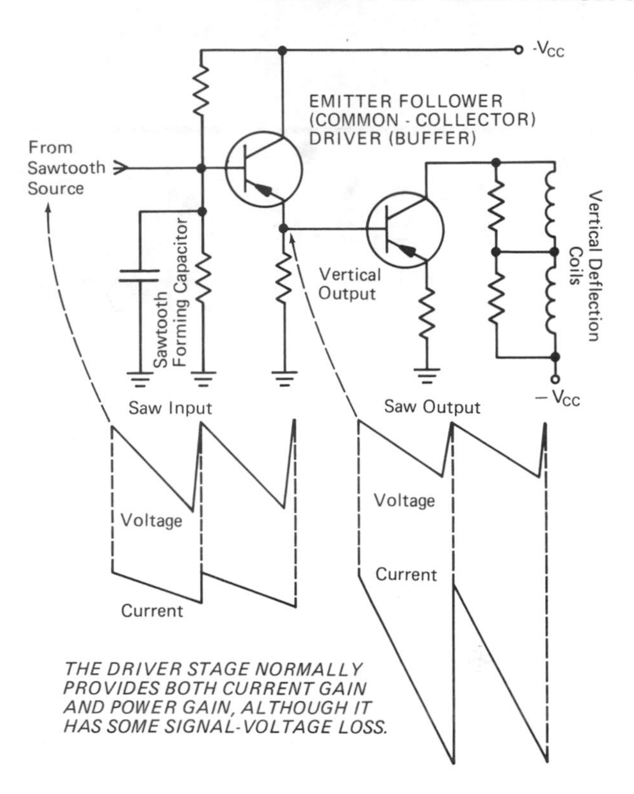

To maintain good sweep linearity, the driver (buffer) stage must be able to supply substantial sawtooth current to the vertical-output stage, and the driver must also draw very little sawtooth current from the sawtooth source. In turn, we often find an emitter follower (common collector) driver arrangement.

Direct coupling is generally employed between the driver transistor and the vertical-output transistor. (However, the vertical-output transistor is less likely to be direct-coupled to the deflection coils, as explained in Chapter 9.)

In normal operation, the emitter follower provides somewhat less output voltage than its input voltage. On the other hand, the emitter follower provides much more output current than its input current. The emitter follower functions as a current amplifier, and in so doing it provides an impedance match (and maximum power transfer) between the sawtooth source circuit and the vertical-output circuit. The emitter follower normally operates in class A.

Preliminary troubleshooting may start with a comparison temperature check of the transistors to determine whether they are running too hot or too cold.

Figure 10-1 Vertical-output stage is commonly driven by an emitter follower. Emitter follower matches high-source Z to low-load Z.

CHART 10-1

Point-to-Point Impedance Checker

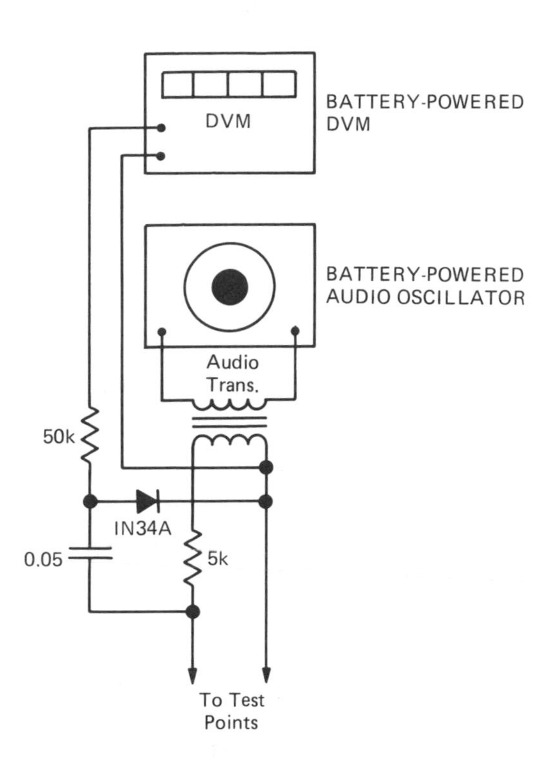

This comparison impedance checker employs a battery-powered voltmeter and battery-powered audio oscillator to minimize the residual capacitance between the test equipment and ground. It is good practice to unplug the receiver under test from the power outlet while making impedance tests between two circuit points, neither of which is connected to chassis ground.

Note that this checker is equally useful for testing impedances to ground.

The frequency of test is not critical, and the troubleshooter chooses a frequency that provides a useful reading on the DVM. In some situations, comparison tests are more informative when the test is repeated at a higher or lower frequency.

As previously noted in discussions of in-circuit impedance testing, the applied test voltage should be restricted to a value below the turn-on voltage of transistors or diodes that are

CHART 10-1 CONTINUED

included in the circuit under test. In other words, *the peak applied test voltage should not exceed 500 mV.* (The DVM indicates the peak applied test voltage.)

An impedance checker provides helpful follow-up test data after low-power resistance measurements have been made, because an ohmmeter cannot indicate whether a capacitor may have lost a substantial proportion of its capacitance, or if it may be open, or if it may have a poor power factor. On the other hand, an impedance checker can usually pick up these faults.

Note that it is poor practice to use an applied test voltage that turns on a transistor or diode junction because junction resistances have a wide tolerance and are also nonlinear. This is just another way of saying that if the applied test voltage is high enough to turn on semiconductor junctions, the DVM readings are difficult to evaluate.

An oscilloscope is also helpful in comparison checks of buffer-stage faults, when waveforms are present. Note that a TV analyzer can be used to inject sawtooth waveforms into a dead stage. In turn, a raster may then be displayed on the picture-tube screen, or an oscilloscope can be used to check for an output waveform from the buffer stage.

If an oscilloscope is not available, a DVM can be used to check for the presence of waveforms, and to measure their peak-to-peak voltages. If your DVM does not have a peak-to-peak voltage function, you can use a peak-to-peak reading probe, as shown in Figure 4-3. Although comparison voltage measurements are very helpful, it is also desirable to know whether the waveshapes are approximately the same, and whether their frequencies are the same. Here is a useful trick of the trade:

With reference to Figure 2-1, a miniamplifier/speaker can be operated as an aural signal tracer, and it provides waveform evaluation also: if the waveshapes of the compared waveforms are substantially different, the *timbre* of the sound outputs will also be different. For example, the sound of a 60-Hz sawtooth wave is distinctly different from the sound of a 60-Hz sine wave, or hum. If the frequencies of the compared waveforms are different, the *pitch* of the sound output will also differ. (See Figure 10-2.)

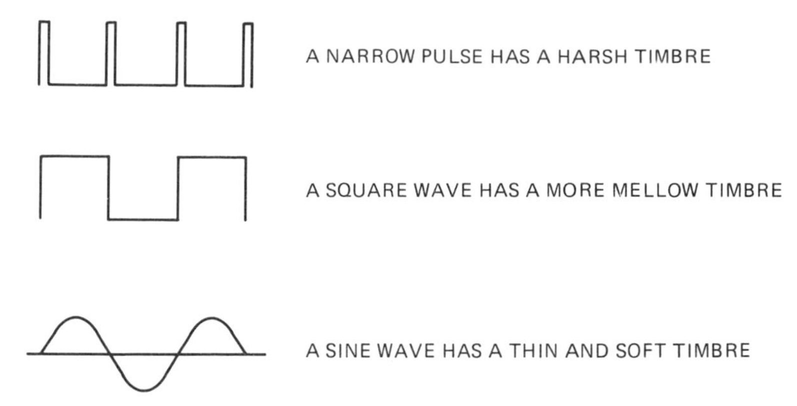

All three of the foregoing waveforms have the same pitch. Their pitch is determined by their repetition rate.

An experienced troubleshooter can tell quite a bit about a waveform from the sound that it makes through an aural signal tracer. The signal tracer should have uniform frequency response from 20 Hz to 20 kHz, and less than 1 percent distortion. Otherwise, a critical evaluation of timbre cannot be made. (The speaker used with the aural signal tracer should also have high-fidelity characteristics.) The best evaluation can be made at moderately loud output level—low-level and very high-level outputs cannot be accurately evaluated.

Follow-up waveform analysis by ear can be made with a processor. For example, a differentiating circuit is a basic processor. An integrating circuit is another basic processor. Illustration: When a true sine wave is differentiated or integrated, its timbre is unchanged. *A distorted sine wave will undergo a change in timbre.*

Figure 10-2 Basic waveforms with distinctive timbres.

COMMON-EMITTER BUFFER STAGE

Another vertical-sweep buffer arrangement that will be encountered is the common-emitter configuration depicted in Figure 10-3. In normal operation the vertical-sweep action is practically linear. Vertical nonlinearity may originate in the sawtooth source, the driver stage, or the output stage.

Nonlinearity that originates in the driver or output stage usually results from leakage, either in a capacitor or in a transistor junction. However, if leakage is not the cause, as determined by DC voltage and resistance checks, the probable cause is an electrolytic capacitor

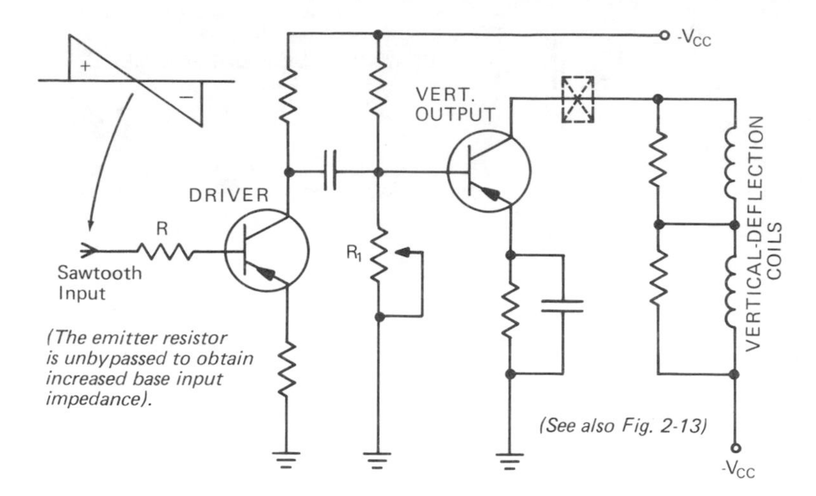

The common-emitter buffer (driver) employs a series input resistance R to reduce the loading on the sawtooth input source. It also utilizes capacitive coupling to the vertical-output transistor. Note that the vertical-output transistor may be direct-coupled to the vertical-deflection coils, although it is more likely to include means for reducing or eliminating flow of the DC component through the deflection coils.

Note that R_1 is a linearity control; it functions to change the transfer characteristic of the vertical-output transistor as required.

This symbol denotes the probable inclusion of elaborated circuitry for reducing or eliminating the flow of DC through the coils.

Note also that the configuration may be further elaborated to return the vertical-deflection coils to ground through the emitter resistor in the driver stage. When the vertical-deflection current flows through the emitter resistor, negative feedback occurs. This feedback provides improved operating stability and linearization.

Figure 10-3 The vertical driver may operate in the CE mode. A CE driver provides comparatively high power gain.

that has lost a substantial portion of its capacitance, or that has a poor power factor. (Collector junction leakage, for example, shifts the bias point, as does a leaky coupling or bypass capacitor.) Subnormal capacitance, or a poor power factor, introduces spurious reactance, with resulting waveform distortion.

VERTICAL OSCILLATOR OPERATION

When troubleshooting the vertical-sweep section, it is helpful to recognize the basic types of vertical oscillator circuitry. Most vertical

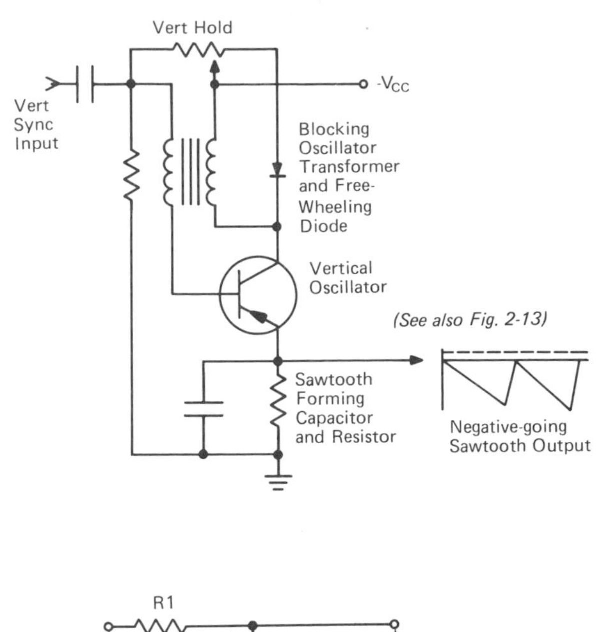

In the free-running state, this blocking oscillator starts with a negative voltage at the base (from the vertical hold control). In turn, the transistor conducts and positive feedback from the transformer drives the transistor into saturation, with heavy current flow. Consequently, the sawtooth capacitor charges up negatively, and the transistor is unable to conduct further. The base voltage drops and cuts off the transistor. In turn, the saw capacitor discharges via the emitter resistor. As the emitter voltage approaches the base voltage, the transistor comes out of cut off, and the oscillatory cycle repeats.

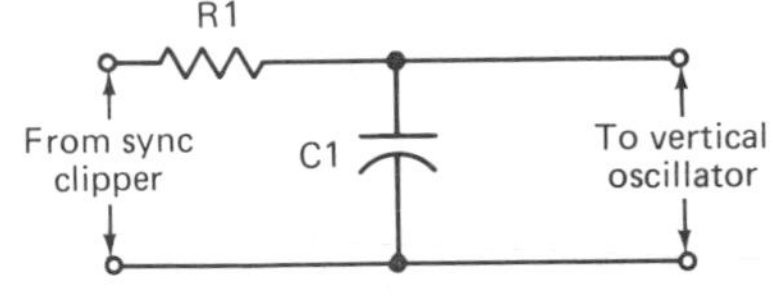

In the synchronized state, the transistor comes out of cut off slightly earlier than in its free-running state, due to application of an integrated vertical-sync pulse to the base of the transistor.

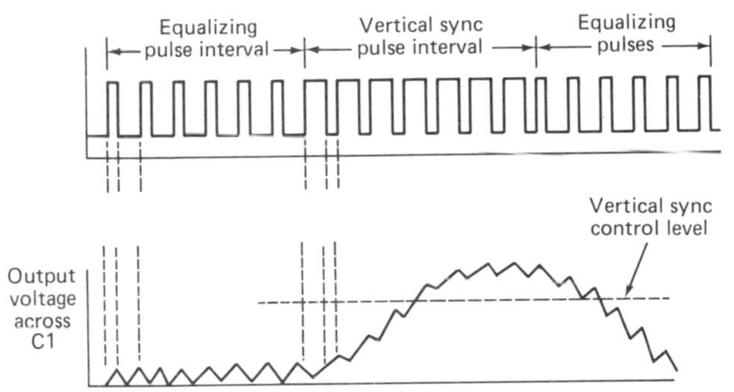

The integrating circuit functions as a low-pass filter. It rejects horizontal sync pulses and equalizing pulses. The vertical trigger pulse develops from the wider pulses within the vertical sync-pulse interval, as shown in these expanded waveform diagrams.

Reproduced by special permission of Reston Publishing Co. and Robert Russell from Electronic Troubleshooting with the Oscilloscope.

Figure 10-4 Basic vertical blocking-oscillator arrangement.

oscillators are designed as blocking oscillators (see Figure 10-4). The transistor functions essentially as an electronic switch, and closes for a short time (retrace interval) 60 times a second.

Note that the free-wheeling diode connected across the primary

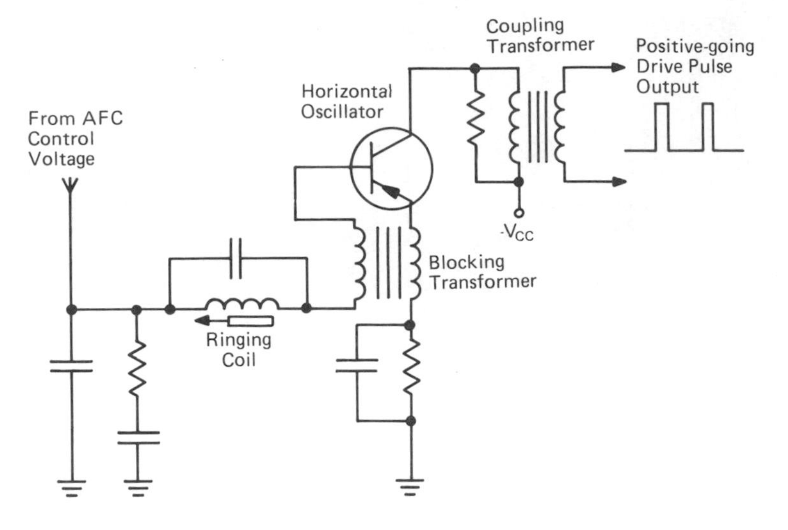

The blocking oscillator normally operates at 15,750 Hz. Note that the RC time constant in the emitter circuit establishes the free-running frequency of the oscillator, and that its exact controlled frequency is determined by the AFC control voltage to the base of the transistor. A ringing coil in series with the base circuit generates a sine wave that rides on the DC voltage at the base of the transistor. The phase of this ringing voltage is determined by the slug adjustment. It is set to bring the transistor out of cut off rapidly, and to thereby minimize the possibility of false triggering by noise pulses. Note that the transistor conducts only during the brief retrace interval. The AFC control voltage is applied through an RC low-pass filter that reduces the loop gain around the AFC section at high frequencies, thereby preventing spurious oscillation. The system operates at high gain at low frequencies, thereby optimizing the phase accuracy in the frequency control process. Faulty capacitors (or off-value resistors) can cause spurious oscillation of the system—the frequency of spurious oscillation can fall in the audible range, or it may fall in the ultrasonic range.

The ringing coil can be identified with a resonance probe—it has a resonant frequency of 15,750 Hz, approximately.

Figure 10-5 Sine-wave stabilized oscillator configuration. A widely used horizontal blocking oscillator arrangement.

of the blocking oscillator transformer effectively short-circuits the primary winding for the kickback pulse (counter EMF) that is generated immediately following the pulse of heavy current flow (retrace interval). If this inductive surge is not suppressed, the vertical-oscillator transistor could be damaged.

A free-wheeling diode is defined as a rectifier diode connected across an inductive load to carry a current proportional to the energy stored in the inductor. The diode conducts this current when zero power is being supplied by the source to the inductive load. The diode conducts until all the energy stored in the inductor has been dissipated (or until the next voltage pulse is applied).

HORIZONTAL BLOCKING OSCILLATOR

We will usually also find blocking-oscillator circuitry employed in the horizontal-sweep section. A basic arrangement is shown in Figure 10-5. An oscilloscope is generally helpful in buzzing out the circuitry. For example, a scope quickly identifies a horizontal circuit versus a vertical circuit (provided that a stone dead condition is not confronted). As noted previously, comparison tests should be made whenever possible.

Note that an aural signal tracer, like that used to trace vertical circuitry, cannot be used to trace horizontal circuitry because most troubleshooters cannot hear a 15,750 Hz frequency. However, an aural signal tracer can be elaborated as shown in Chart 10-2 to trace horizontal circuitry. It is a heterodyne or beat type of tracer that converts horizontal pulses into audio pulses.

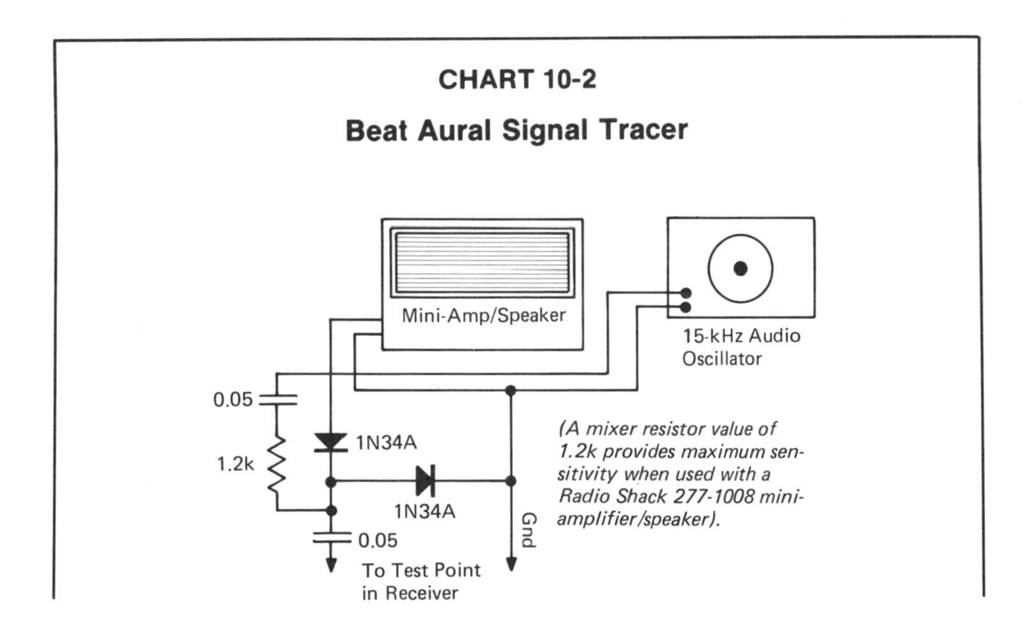

CHART 10-2

Beat Aural Signal Tracer

CHART 10-2 CONTINUED

This is a heterodyne or beat type of aural signal tracer. In this example, the 15-kHz sine-wave voltage from the audio oscillator heterodynes with 15,750-Hz pulses in the 1N34A diodes. In turn, a 750-Hz beat voltage is applied to the miniamp/speaker, which provides a 750-Hz tone output.

The output level from the audio oscillator is adjusted to provide optimum sound output from the speaker. (The volume control on the miniamp is also adjusted for a suitable sound output level.)

When an oscilloscope is not available, this beat aural signal tracer enables the troubleshooter to identify ultrasonic circuits that are beyond the capability of conventional aural signal tracers.

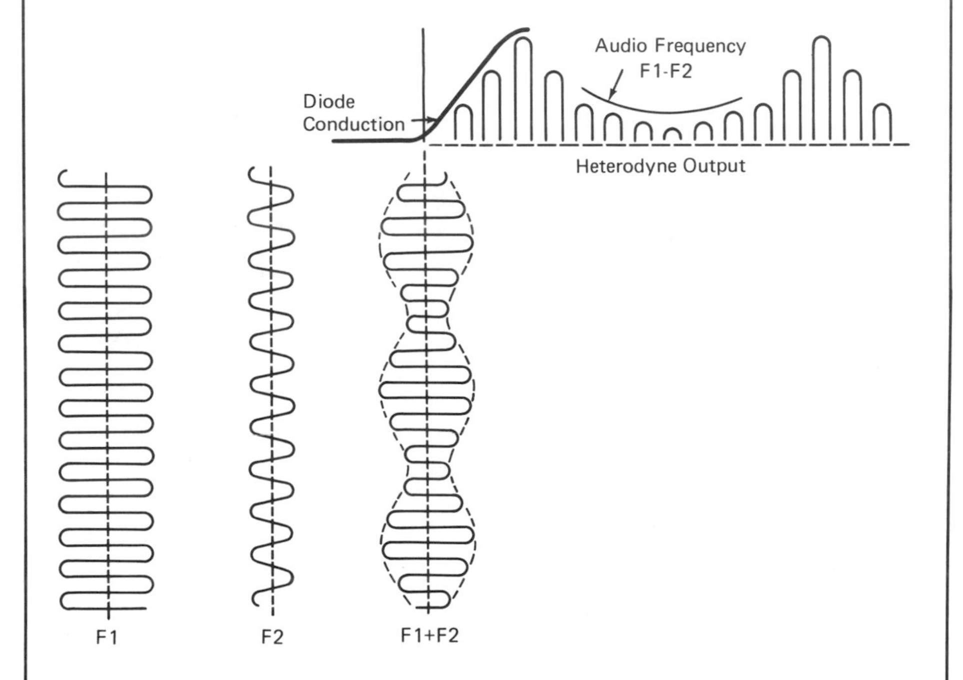

Heterodyne action is used to shift a higher signal frequency to some lower frequency. It operates by forming the sum of two frequencies (mixing the two signals together). This sum waveform is then passed through nonlinear resistance, with the result that the difference between the two frequencies (beat frequency) becomes available. This difference frequency is one of the envelopes of F1 + F2.

HORIZONTAL AFC STAGE

It is helpful to recognize the widely used horizontal AFC arrangements. The transistor type is exemplified in Figure 10-6; it operates as a phase detector and compares the phase of the incoming sync pulses with the phase of a comparison sawtooth from the horizontal-output section. Another AFC arrangement is actually quite similar, except that it employs a pair of diodes, instead of using the emitter and collector junctions in a transistor.

When an oscilloscope is not available, sync-section pulses can be analyzed to a considerable extent by means of probes such as that depicted in Figure 4-3. In other words, pulses may be positive-going or negative-going, or they may have equal positive and negative excursions. For example, if the DVM reading is much higher with a positive-peak probe than with a negative-peak probe, the troubleshooter knows that the circuit under test operates with positive-going pulses.

Conversely, if the DVM reading is much higher with a negative-peak probe than with a positive-peak probe, the troubleshooter knows that the circuit under test operates with negative-going pulses.

Again, if the DVM readings are practically the same with a positive-peak probe and with a negative-peak probe, the troubleshooter knows that the circuit under test operates with symmetrical pulses.

Note also that sync-section waveforms can be further analyzed by comparison of their peak voltages vs. their average voltages. As an illustration, with reference to Figure 10-7, the average voltage of a sawtooth waveform is 25 percent of its peak voltage. It is also a symmetrical waveform—its positive peak and average voltages are the same as its negative peak and average voltages.

To recap an important principle, most DVMs respond to the average value of a half cycle on their AC voltage measuring function. If the test leads to the DVM are reversed, the meter responds to the average value of the opposite-polarity half cycle. However, the readout is generally 2.22 times the average value. In other words:

1. A DVM generally employs a half-wave instrument rectifier on its AC voltage function.
2. The meter responds to the average value of the half-rectified waveform.

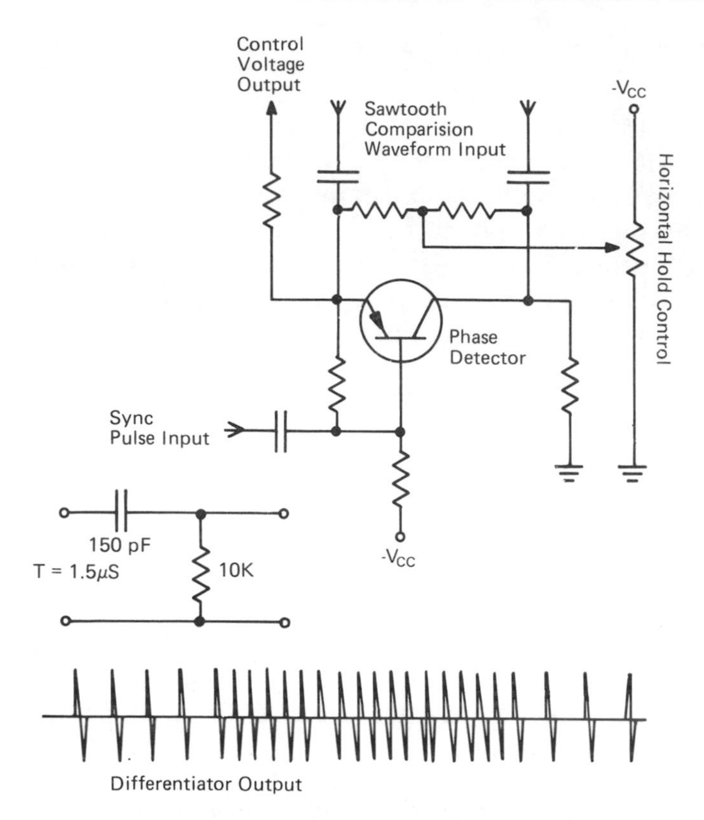

The horizontal phase detector develops a DC control voltage which adds to or subtracts from the fixed bias on the horizontal-oscillator transistor, and thereby corrects any tendency of the horizontal oscillator to drift off frequency.

This control voltage may be either positive or negative. Note that the phase detector transistor bring the transistor out of cut off into conduction. The sync pulse input capacitor is charged at this time by the flow of base current, and holds the transistor in cut off during the interval between sync pulses.

Negative-going differentiated sync pulses applied at the base of the phase detector transistor bring the transistor out of cut-off into conduction. The sync pulse input capacitor is charged at this time by the flow of base current, and holds the transistor in cut-off during the interval between sync pulses.

As noted in the diagram, the collector-to-emitter voltage is a sawtooth waveform obtained from the horizontal-output stage. This collector-to-emitter voltage is negative during the first half of the retrace interval, and the transistor conducts in the usual mode. However, during the latter half of the retrace interval, the collector goes positive and now

Figure 10-6 A widely used horizontal-AFC arrangement.

functions as an emitter. If the horizontal oscillator is properly phased, the center of the sync pulse precisely straddles the point where the sweep signal passes through zero during the retrace interval.

Under this condition, during the time that the sync pulse holds the transistor in conduction, current flows first from emitter to collector; then the current reverses and flows from collector to emitter. Since the peak current amplitude in this situation is the same for both directions of flow, the resulting control voltage output is zero.

However, if the horizontal oscillator starts to pull, the sync pulse will arrive a bit early, or a bit late with respect to the point where the sweep signal passes through zero during the retrace interval. In turn, the current flow through the phase detector is greater in one direction and less in the other direction. Consequently, a positive or a negative control-voltage output is produced. This control voltage functions to bring the horizontal oscillator back on frequency.

Figure 10-6 (continued)

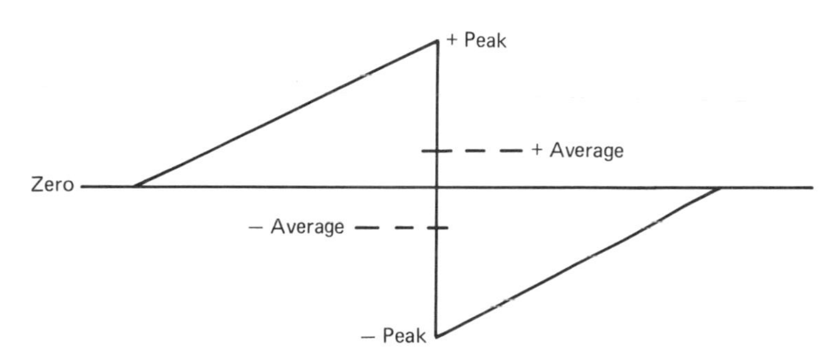

Note: This is a symmetrical waveform because the positive and negative excursions have the same shape.

Its + average value is 0.25 of peak, and its − average value is also 0.25 of peak.

When making comparison tests, waveform distortion can often be picked up by measuring both positive and negative peak voltages, and positive and negative average voltages. In other words, the ratios of peak to average voltages are waveshape indicators. As a basic illustration, a sine wave has a peak-to-average voltage ratio of 3.14, whereas a sawtooth wave has a peak-to-average voltage ratio of 4.00.

If you are concerned with true average voltages, remember that most DVMs read 2.22 times the true average voltage of a half-rectified waveform.

Figure 10-7 Peak and average values in an AC sawtooth waveform.

3. The instrument is calibrated to indicate the r.m.s. value of a sine waveform.
4. Therefore, the readout is equal to 2.22 times the average value of the waveform (0.707/0.318 = 2.22).

In preliminary troubleshooting procedures, the technician is not concerned with the actual average values of complex waveforms. However, he is often concerned with relative values. For example:

1. A waveform measures 2 volts positive peak, and 2 volts negative peak.
2. The same waveform measures 1.11 volts positive average, and 1.11 volts negative average value.
3. In turn, the troubleshooter knows that the waveform is symmetrical.
4. With reference to Figure 10-7 (and taking the 2.22 readout factor into account) the troubleshooter would conclude that the waveform is very likely to be a sawtooth.

HORIZONTAL DRIVER STAGE

It is also helpful to recognize the widely used horizontal-driver arrangements. A horizontal-driver transistor usually operates in class B, as exemplified in Figure 10-8. In other words, if there is no input waveform from the horizontal oscillator, the base-emitter bias on the driver transistor is zero.

As noted in the diagram, an adjustable amount of back bias is normally present on the driver transistor when an input waveform from the horizontal oscillator is present. This back bias provides marginal class-C operation of the driver transistor and avoids driving too far into saturation. The amount of back-bias voltage that is developed depends on the RC time constant of the bias leg in the base-emitter circuit.

If the total current flow through the driver transistor is too large, its temperature will be abnormal. On the other hand, if the total current flow is too small, its temperature will be subnormal. Follow-up tests include comparative DC voltage and resistance checks, and impedance checks. Oscilloscope waveform checks can often show whether a trouble symptom is being caused by improper drive, by an input circuit fault, or by an output circuit fault.

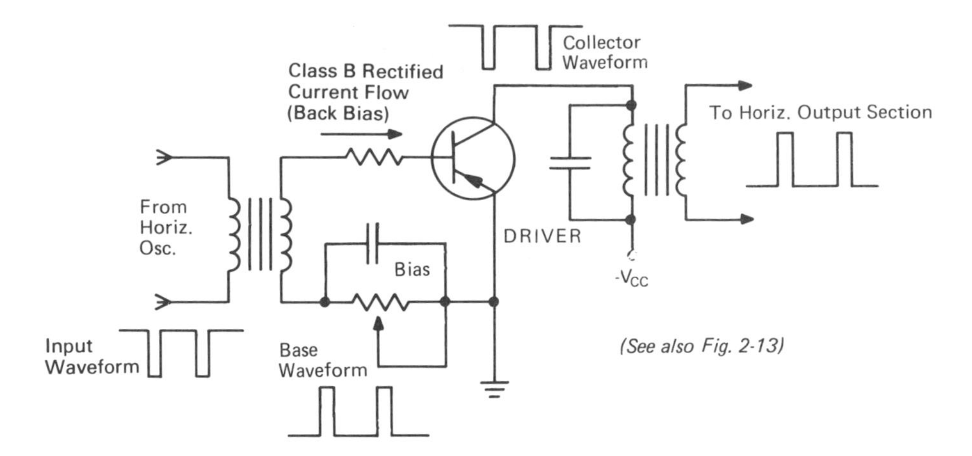

Preliminary troubleshooting should start with a comparison temperature check of the driver transistor. (It might be running too hot or too cold.)

The driver stage inputs and outputs a pulse waveform. Note that the input signal from the horizontal oscillator holds the driver transistor in conduction (saturation) during the forward-scanning interval, and cuts off the transistor during the retrace interval. The variable resistor in the base circuit is adjusted so that a small amount of signal-developed reverse bias is provided.

An excessive amount of signal-developed reverse bias should be avoided, because the amplitude of the output waveform would become subnormal. On the other hand, an insufficient amount of signal-developed reverse bias should be avoided, inasmuch as the output transistor would be cut off too slowly, with the result that overheating could occur.

The fixed resistor in the base circuit functions as a current limiter, to assist in keeping the driver transistor out of deep saturation.

Figure 10-8 Typical configuration for a horizontal driver stage. Horizontal driver operates as a buffer and power amplifier.

HORIZONTAL OUTPUT AND HIGH VOLTAGE SECTION

Troubleshooting the horizontal-output and high-voltage section (Figure 10-9) is a comparatively demanding task, chiefly because the network performs several functions. It is not only a deflection arrangement, but it is also an energy-recovery system and a high-voltage supply source.

The horizontal-output transistor operates as a switch—it alternates between cut off and saturation. A basic requirement to prevent

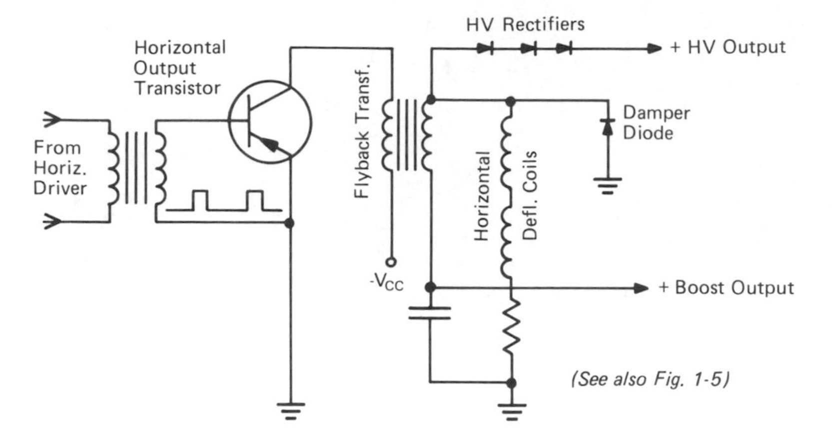

Preliminary troubleshooting should start with a comparison temperature check of the horizontal-output transistor. (It might be running too hot or too cold.)

The horizontal-output transistor normally operates in class B. In other words, if there is no input pulse from the horizontal driver, the base-emitter bias is zero. The horizontal-output transistor outputs an essentially rectangular voltage waveform for driving sawtooth current through the horizontal deflection coils. Flyback (retrace) occurs during a brief interval when the horizontal output transistor is suddenly cut off by a pulse from the driver. The flyback transformer and deflection coils, in turn, "ring" for one-half cycle of a 70-kHz sine wave. Ringing normally stops at the end of a half cycle, due to damper conduction.

The damper diode normally conducts for approximately the first half of each forward-scan interval, whereupon the horizontal-output transistor comes out of cut off, goes into conduction and into saturation.

A TV analyzer can be used to inject drive pulses into the horizontal output circuit.

Figure 10-9 Skeleton horizontal-output and high-voltage section. Output circuitry provides deflection, boost, and high voltages.

overheating of the horizontal-output transistor is a drive pulse with a fast rise. As noted above, an essential condition for fast rise is correct bias voltage on the driver transistor, to avoid deep saturation and resulting charge storage in the transistor.

Note that if the high-voltage rectifier diodes break down, the picture-tube screen will be dark, although the horizontal sweep action continues as can be determined by scope waveform checks.

Sweep action is also evidenced in this trouble situation by the presence of normal boost voltage output.

It is sometimes overlooked that the horizontal-output transistor must conduct current from collector to emitter, and also from emitter to collector. In other words, if the transistor is not driven sufficiently into saturation, it will not be able to conduct current over the complete forward-scan interval.

In other words, the horizontal-output transistor conducts in the usual manner during the second half of the forward-scan interval. At the end of forward scan, there is considerable magnetic energy stored in the deflection coils—the transistor is suddenly cut off by the drive pulse, and retrace ensues with reversal of voltage across the deflection coils due to counter e.m.f. In turn:

The stored energy in the deflection coils proceeds to force current flow through the flyback transformer in the opposite direction, thereby providing the first half of the following forward-scan interval. Since the transistor is normally in saturation, its collector now functions as an emitter, and its emitter now functions as a collector.

Thus, the first half of the forward-scan interval occurs normally, provided that the transistor is maintained in saturation during this time by the drive waveform. At the end of the first half of the forward-scan interval, the stored energy in the deflection coils has decayed to zero, and the transistor stops conducting reverse current. Now, the drive waveform takes over and the transistor starts conducting electrons from collector to emitter—the sawtooth deflection current continues to build up until the next drive pulse suddenly cuts off the transistor, and the cycle repeats.

As a rough rule of thumb, it is frequently stated that horizontal-sweep distortion in the left-hand region of the picture-tube screen indicates damper-circuit defects, whereas distortion in the right-hand region of the screen indicates drive-circuit defects. Of course, a flyback-transformer fault, or a deflection-coil fault will affect the entire deflection cycle.

Comparison Temperature Checking

Comparison temperature checks can sometimes narrow down the trouble area rapidly. For example, if the driver transistor is running hot, and the horizontal-output transistor is running too cool, the troubleshooter will turn his attention to the driver section. Note that when an identical receiver is not available for comparison tests, a somewhat similar receiver may provide helpful test data. Transistors rated for the same power output will normally operate at approximately the same temperature. Again, the same type of transistor used in different types of horizontal-output circuitry will normally operate at about the same temperature in both circuits.

Power transistors mounted on heat sinks must make good thermal contact with metal surfaces. In case of doubt, coat the surfaces with silicone grease and secure the transistor tightly in place. If the transistor then runs cooler, the troubleshooter knows that heat-sink trouble was present.

BACKGROUND SHADING TROUBLE SYMPTOMS

The power transistors used in TV receivers has a comparatively heavy current demand which often varies considerably over the forward-scan interval. In other words, one or more power transistors may draw a relatively small current over the first quarter of the forward-scan interval, and then suddenly draw a considerably greater amount of current that gradually diminishes over the remainder of the forward-scan interval.

In turn, a varying current demand is placed on the power supply over the forward-scan interval, and, if the power-supply regulation is inadequate, there will be an abnormally high ripple on the V_{CC} line. The regulation is often determined entirely by the condition of the filter capacitors and decoupling capacitors in the receiver. Accordingly, if a scope check at the picture-tube terminals shows an abnormally high ripple amplitude, look for deteriorated electrolytic capacitors in the power supply and at the decoupling points along the V_{CC} line.

11

Color Television Troubleshooting

*Preliminary Trouble Analysis * Chroma Circuit Identification * Chroma Signal Levels * Specialized Chroma Test Signals * Voltage and Resistance Measurements * Reverse-Polarity Base Bias * Troubleshooting Integrated Circuitry * Color-Killer Voltage and Current * Bandpass Amplifier Troubleshooting * Color Sync Gate and Amplifier * Color Demodulator Troubleshooting*

PRELIMINARY TROUBLE ANALYSIS

Preliminary trouble analysis of a malfunctioning color-TV receiver is made as summarized in Chart 11-1. Color receivers have the same circuit sections that were discussed for black-and-white receivers in Chapter 7, plus chroma (color processing) circuitry, color-sync circuits, and picture-tube convergence configurations. Although there are more functional sections for the troubleshooter to contend with, the same basic electrical and electronic principles apply. (See Chart 11-2.)

From a generalized point of view, the chroma section in a color receiver can be regarded as an elaboration of the video amplifier section. One distinctive characteristic that the technician encounters in the chroma section is the processing of three-phase 3.58-MHz signal voltages. Nevertheless, this elaboration involves only an extension of fundamental video-signal characteristics. The color-sync section is more analogous to the horizontal-AFC section in a black-and-white receiver, than to sync-clipper and separator sections.

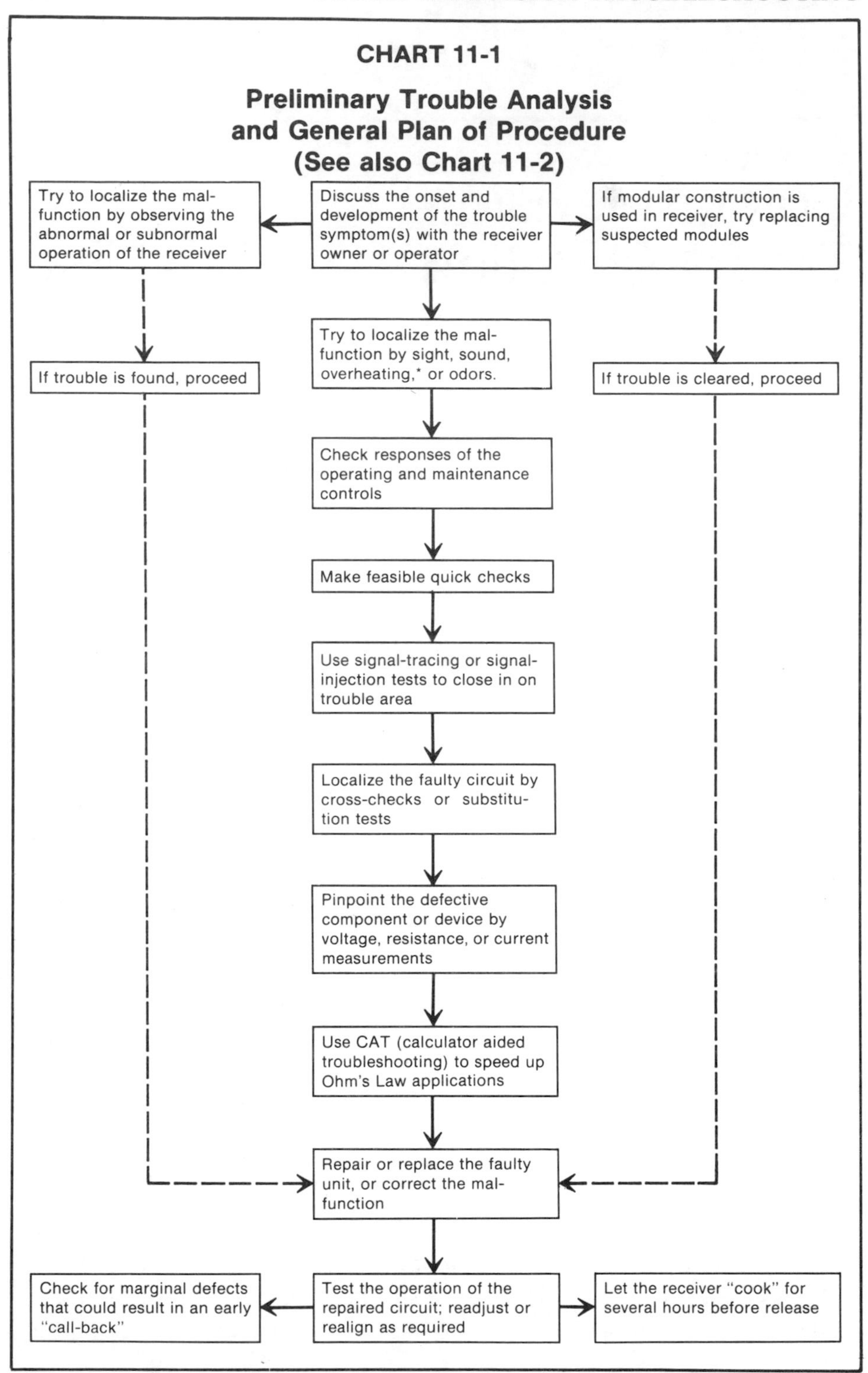

*Temperatures can be quickly and accurately measured with a temperature probe and DVM

CHART 11-2

Chroma Circuit Identification

When troubleshooting without service data, chroma circuit identification can be facilitated by:

1. Oscilloscope waveform checks in live circuitry.
2. Resonance probe checks in dead circuitry. (Figure 4-3.)

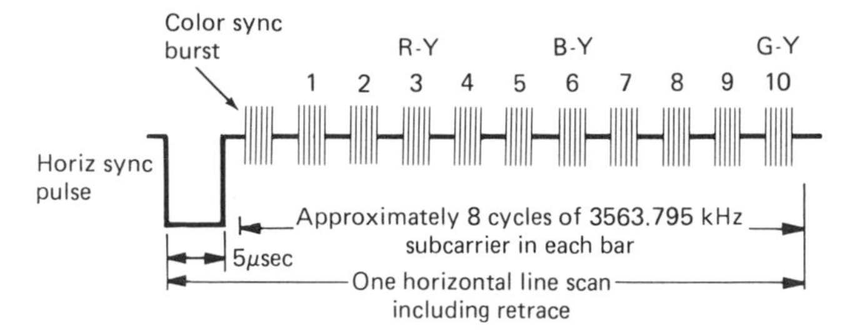

A keyed-rainbow signal should be used. It has a horizontal sync pulse and 11 "bursts."

The first burst serves for color sync action. Ten color bars are normally displayed on the picture-tube screen.

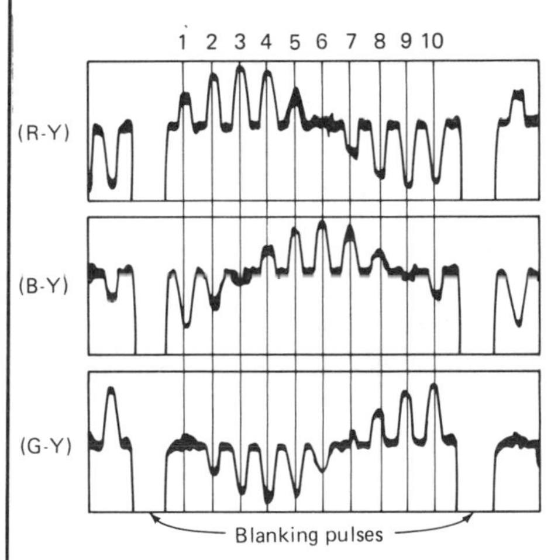

These waveforms are normally displayed at the outputs of the chroma demodulators and chroma matrices.

The oscilloscope should be used with a low-capacitance probe to minimize circuit loading.

Note that XZ chroma demodulators, or RGB matrices may be encountered. However, the general type of waveform is the same—only the zero-crossing points will differ. (In all types of color receivers, the keyed-rainbow signal will normally be displayed at the output of the chroma bandpass amplifier.)

A resonance probe identifies burst-amplifier and bandpass amplifier stages on the basis of their 3.58-MHz resonant-frequency response.

If you do not have an oscilloscope available, you can identify chroma circuitry with an aural signal tracer, depicted in Figure

CHART 11-2 CONTINUED

7-4. The only difference is that a 3.58-MHz frequency source is used (instead of a 4.5-MHz source). The test signal will heterodyne against a color-TV station signal, and also with the subcarrier-oscillator signal in the receiver. You can also use a short-wave radio receiver for a chroma signal tracer, as depicted in Figure 7-5, with the generator tuned to approximately 3.58 MHz.

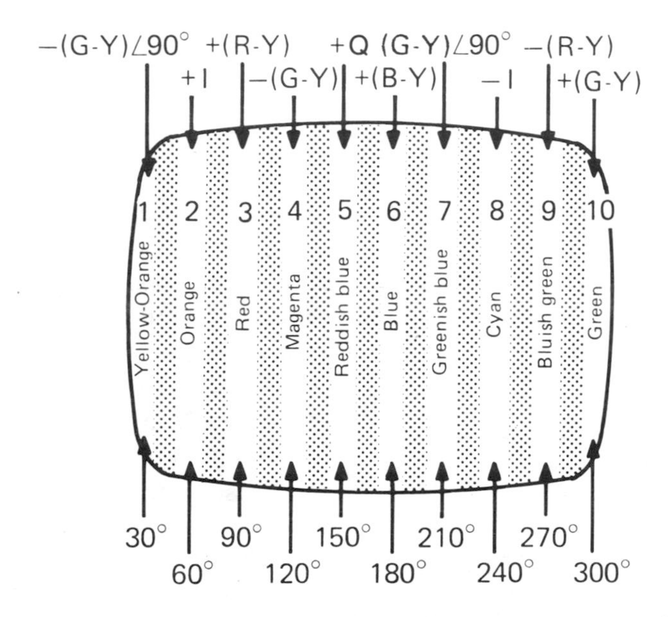

Weak red reproduction points to trouble in the R-Y circuitry. Weak blue reproduction points to trouble in the B-Y circuitry. Weak green reproduction points to trouble in the G-Y circuitry.

The possibility of picture-tube trouble can be checked with a color picture tube test jig.

Lack of color reproduction (black-and-white only) can be caused by a defect in the color-killer section, bandpass amplifier section, or subcarrier-oscillator section.

Reproduced by special permission of Reston Publishing Co. and Miles Ritter-Sanders, Jr. from Handbook of Oscilloscope Waveform Analysis.

Figure 11-1 Color-bar pattern produced on picture-tube screen by the keyed-rainbow signal shown in Chart 11-2.

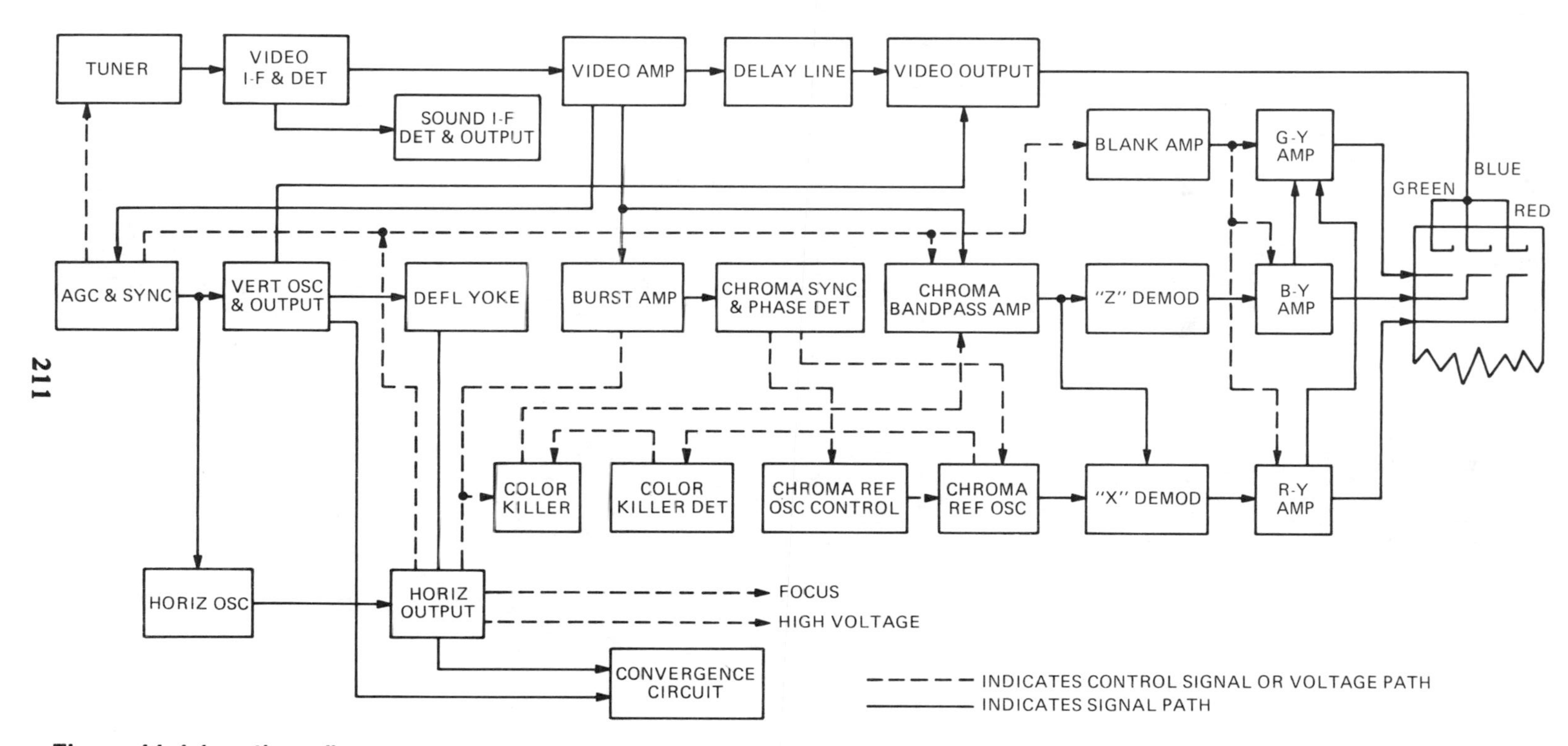

Figure 11-1 (continued)

CHROMA SIGNAL LEVELS

The 3.58-MHz chroma signal is processed at a comparatively low level in modern color-TV receivers. For this reason, *it is not practical to use conventional peak-reading probes and voltmeters to trace the chroma signal.* Even a forward-biased peak-reading probe has limited usefulness in chroma-circuit troubleshooting. Instead, the technician generally uses chroma signal-injection tests to close in on a trouble area. The color picture tube is ordinarily used as an indicator. (See Figure 11-1.) However, if the screen is dark, *the troubleshooter can use a peak-reading probe and a DVM as an indicator at the picture tube cathode or at a grid.*

Specialized Chroma Test Signals

Although a conventional AM generator has considerable utility in preliminary tests of chroma circuitry, evaluation of signal processing action is greatly facilitated by the use of specialized test signals. The most common chroma test signal is provided by a keyed-rainbow generator. This signal waveform comprises groups (bursts) of 3.56-MHz sine-wave voltage; 11 bursts are contained between consecutive horizontal sync pulses.

A keyed-rainbow signal is also called a sidelock signal, an offset color subcarrier signal, or a linear phase sweep signal. Note that the frequency of a rainbow signal is customarily 15,750 Hz less than the color-subcarrier frequency (3.58 MHz). An analyzer generator provides a keyed-rainbow signal at VHF, IF, and video frequencies, convergence test patterns, various pulse outputs, a multiburst frequency-check pattern, and sound-channel test signals. These specialized chroma signals are used in signal-injection tests.

VOLTAGE AND RESISTANCE MEASUREMENTS

DC voltage measurements in the chroma section may be made with or without signal present. Standard service data generally specifies DC voltage values with a keyed-rainbow signal present. In turn, a shift in DC voltage distribution that is not caused by a component or a device defect indicates that the signal voltage is not normal, or is absent. As an illustration, a burst amplifier circuit in a

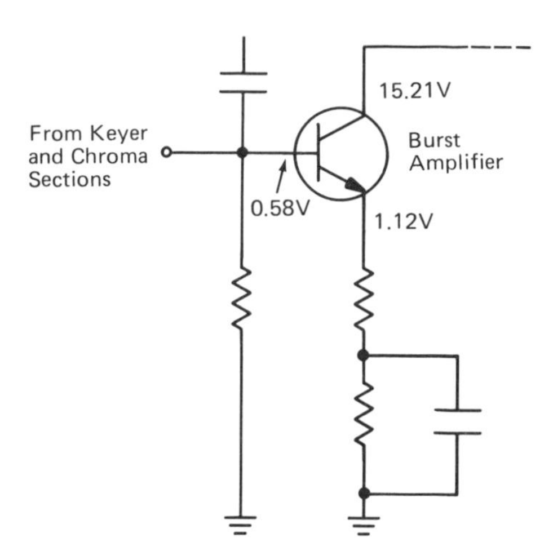

Comparative impedance checks with the arrangement depicted in Figure 9-4 can provide useful data to supplement DC voltage and resistance measurements.

Note: The burst amplifier transistor is normally reverse-biased by 0.54 volt because it is driven into saturation by the keyer pulse and in turn acquires signal-developed bias.

Figure 11-2 Burst-amplifier transistor typically operates in class C and is reverse-biased if the keyer signal is present.

Panasonic color-TV receiver is shown in Figure 11-2. Note the following relations:

1. The transistor is *apparently* cut off, inasmuch as the emitter is more positive than the base.
2. However, the transistor is *not* cut off, since the emitter terminal is considerably above ground potential.
3. Because the base is driven by a strong positive pulse, the transistor conducts heavily on the peak of the pulse, and the emitter capacitor is charged to an average value of 1.12 volts.
4. This capacitor charge is signal-developed bias, and its value is sufficient to make the transistor operate in class C.
5. Therefore, the DC voltage distribution indicates that normal signal voltage is present. If the troubleshooter measured zero volts at the emitter terminal, he or she would conclude either that the keyer signal voltage is not present, or that the transistor is short-circuited.

Reverse-Polarity Base Bias

Another example of DC voltage distribution in a burst amplifier with signal present is shown in Figure 11-3. This is a configuration used in a JVC color-TV receiver. Note the following relations:

1. The transistor is *apparently* cut off, and the base bias voltage has *opposite polarity* from the emitter bias voltage.
2. However, the transistor is *not* cut off, since the emitter terminal is considerably above ground potential.
3. Because the base is driven by a strong positive pulse, the transistor conducts heavily on the peak of the pulse. In turn, the emitter capacitor is charged to an average value of 1.6 volts, and the capacitors in the base circuit are charged to an average value of −0.08 volt.
4. The transistor is substantially reverse-biased by the signal-developed bias voltage, with the result that it operates in class C.
5. Therefore, the DC voltage distribution indicates that normal signal voltage is present. If the troubleshooter measured zero volts at the emitter terminal, he would conclude either that the keyer signal voltage is not present, or that the transistor is short-circuited.

Note in passing that the total signal voltage on the base of the burst-amplifier transistor is a comparatively small-amplitude 3.58-MHz burst, and a comparatively large positive pulse. The burst signal originates in the keyed-rainbow generator that is connected to the receiver, and the keyer pulse is an internally generated waveform from the horizontal-deflection section.

TROUBLESHOOTING INTEGRATED CIRCUITRY

The technician will encounter integrated circuitry (exemplified in Figure 11-4) occasionally. From a practical viewpoint, an integrated circuit is an elaborated transistor—it is characterized by DC voltages and current values. Also, the semiconductor junctions in an integrated circuit are the same as transistor junctions. In-circuit measurements of resistance values can be made with a low-power

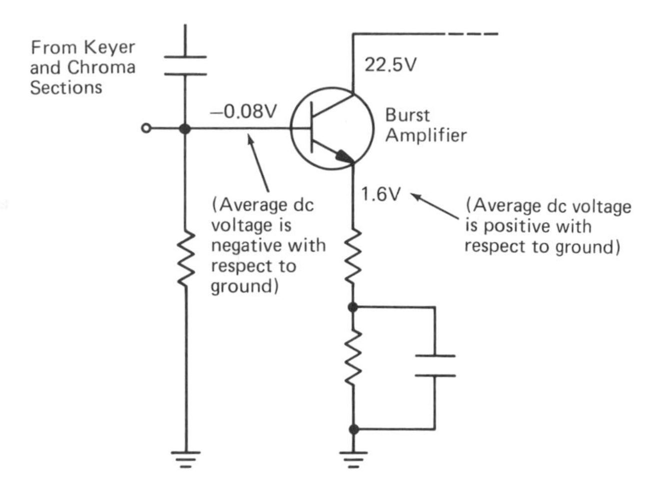

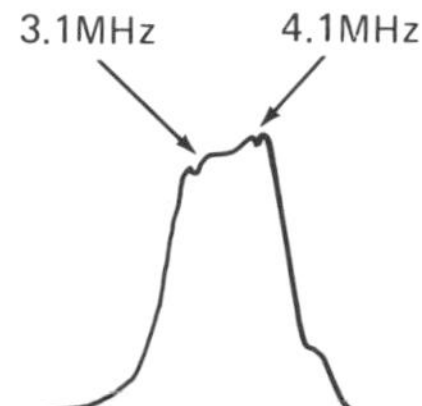

A check of the burst-amplifier frequency response curve can also provide useful data to supplement DC voltage and resistance measurements.

Note: The burst-amplifier transistor is pulsed positively, but the signal-developed bias on the base is negative because of the heavy base-current flow on the peaks of the keyer pulse. This pulsed base current charges the capacitors in the base circuit to an average value of −0.08 volt.

Figure 11-3 Keyer pulse waveform can be checked with an oscilloscope. Another example of a reverse-biased burst-amplifier transistor.

ohmmeter in the same manner as previously explained for transistor circuits.

Note that the DC voltages specified for an integrated chroma section may be "mixed," in that some of the voltages may be specified with chroma signal present, whereas other voltages may be specified with chroma signal absent. Check the service data for the particular color receiver to determine whether a given voltage value should be checked with chroma signal present, or with chroma signal absent. The chroma signal is supplied by a keyed-rainbow generator.

The tolerance on specified DC voltage and current values is ±20 percent, unless otherwise noted in the service data. Similarly, the

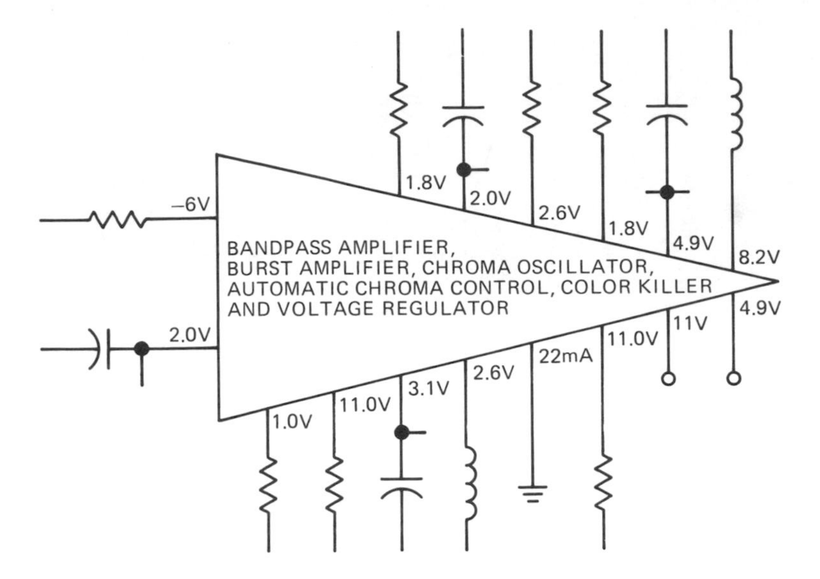

The 3.58-MHz signal terminals can be buzzed out with the aid of a portable short-wave radio receiver and AM generator, using the method shown in Figure 7-5.

Voltage, current, and resistance measurements are most helpful when made on a comparison basis in an identical good receiver. However, these measurements are also a useful guide when made on the same type of IC in a somewhat similar receiver.

Note: The 22-mA current demand can be checked by slitting the PC conductor with a razor blade. After the current is measured with a milliammeter, the slit is closed with a small drop of solder.

If you are not familiar with the integrated circuits in modern color-TV receivers, refer to pages 77 through 109 in *Encyclopedia of Integrated Circuits*, by Walter H. Buchsbaum (Prentice-Hall, 1980).

Figure 11-4 Integrated circuitry is often used in the chroma section. Example of an integrated input chroma section, with specified DC voltage and current values.

tolerance on resistance values in the branch circuits is ordinarily ±20 percent. DC voltages and currents should be measured with the line voltage adjusted to rated value for the receiver (usually 117 V). If a capacitor in a branch circuit is suspected of leakage, one end may be unsoldered for test, or, the PC conductor may be slit with a razor blade. After the test is made, the slit is closed with a small drop of solder.

Resistance Charts

An example of a resistance measurements chart is shown in Figure 11-5. These are in-circuit resistance values, measured with a low-power ohmmeter from device terminals to ground. Note that an incorrect in-circuit resistance value may be caused either by a defective component or a defective device. For example, if there is a leaky capacitor in a transistor branch circuit, the branch-circuit resistance will be off-value. However, if the transistor junction is defective, the branch-circuit resistance will also be off-value.

Therefore, when a resistance value is out of tolerance with the specified value in the resistance measurements chart, additional tests must be made to pinpoint the fault. *In the case of a transistor, each terminal connects internally to a junction, and infinite resistance will normally be measured from any one terminal to another when a low-power ohmmeter is used.*

On the other hand, each terminal of an integrated circuit does not necessarily connect internally to a junction—an IC terminal may

Item	Pin 1	Pin 2	Pin 3	Pin 4	Pin 5	Pin 6	Pin 7
Q1	23 Ω	3 k Ω	69 k Ω				
U1	8 k Ω	17 Ω	41 Ω	47 k Ω	59 Ω	12 k Ω	300 Ω
Q2							
U2							
Q3							
Q4							
Q5							

Note: Transistors are customarily identified as Q1, Q2, etc. Integrated circuits may be identified as U1, U2, etc., or IC1, IC2, etc. All resistance values are specified from device terminals to ground, using a low-power ohmmeter.

When troubleshooting without service data, resistance-to-ground values are most informative when made on a comparison basis with an identical good receiver. However, when an identical receiver is not available, a somewhat similar good receiver that uses the same type of IC can provide useful ball-park resistance-to-ground values.

Resistance charts are published by Howard W. Sams & Co., Inc.

Figure 11-5 Example of a resistance measurements chart.

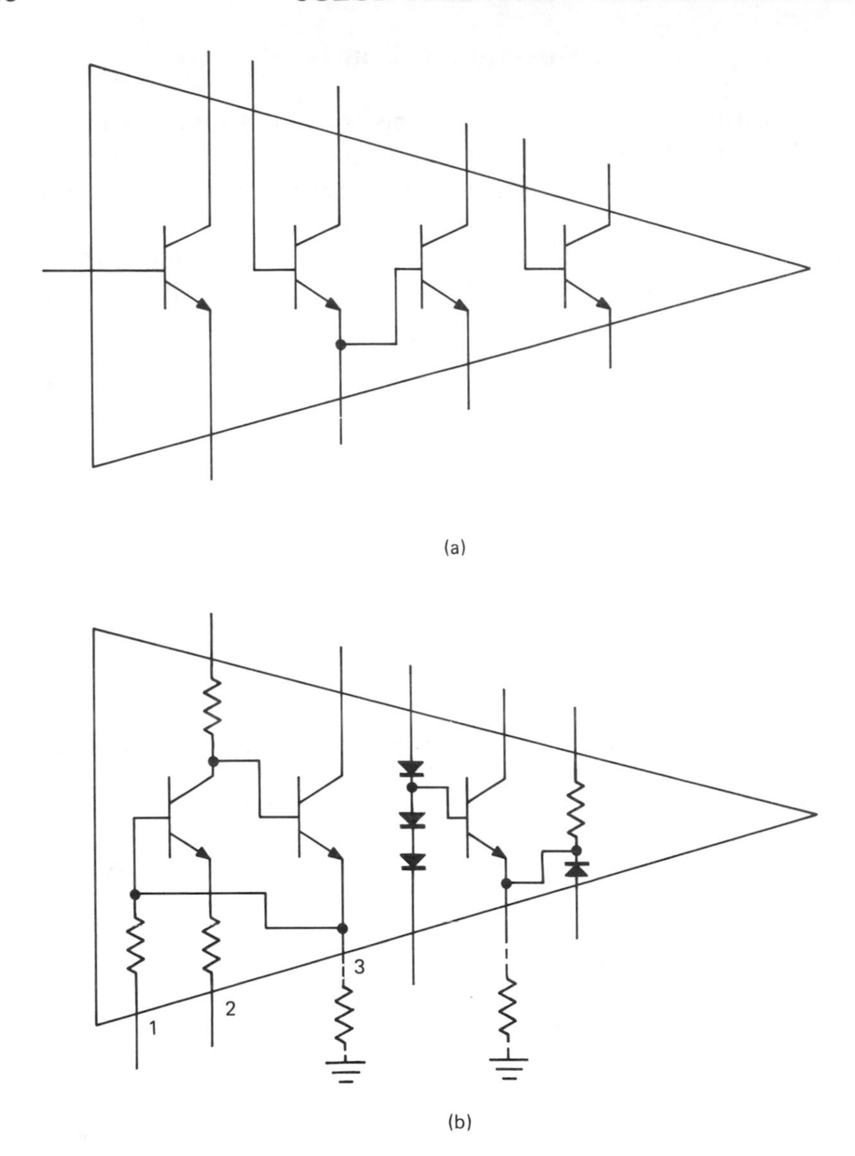

Some ICs contain transistors only—others contain transistors, resistors, and also diodes.

Figure 11-6 Basic examples of integrated circuitry. (a) All IC terminals connect to junctions; (b) some IC terminals connect to others through resistors.

connect internally to a resistor, which in turn is connected to another IC terminal, as shown in Figure 11-6.

When an out-of-tolerance resistance value is measured from an IC terminal to ground with a low-power ohmmeter, a meaningful cross-check can be made in the example of Figure 11-6(a) by slitting the PC conductor to the given IC terminal. Then, if an infinite resistance is measured at the given IC terminal with respect to ground, the troubleshooter concludes that the trouble will be found in the branch circuit, and not in the IC.

However, a meaningful cross-check cannot necessarily be made in this manner in the example of Figure 11-6(b). As an illustration, if the troubleshooter measures an out-of-tolerance resistance value from pin 1 to ground, he could proceed to slit the PC conductor at pin 1. Unfortunately, in this case, the IC will *appear* to be defective because pin 1 has a resistive return to ground via pin 3. Therefore, the following procedure should be observed:

1. When an out-of-tolerance resistance value to ground is measured at an IC terminal with a low-power ohmmeter, slit the PC conductor at the IC terminal.
2. Check out the components and devices in the branch circuit that has been disconnected from the IC.
3. If all the components and devices in the branch circuit check out satisfactorily, it can then be concluded that the IC is defective.

In the vast majority of situations, the troubleshooter will not know whether the IC under test is an all-junction type as in Figure 11-6(a), or a junction-and-resistor type, as in Figure 11-6(b). Therefore, the only practical course is to assume that the IC is a junction-and-resistor type, and to proceed accordingly.

Color Killer Voltage and Current

A color killer configuration operates as an electronic switch to turn off the chroma-signal processing channel during black-and-white reception. The color killer is actuated by the color burst. If the color-killer circuit malfunctions, confetti (colored snow) is displayed in the raster during black-and-white reception. The color-killer

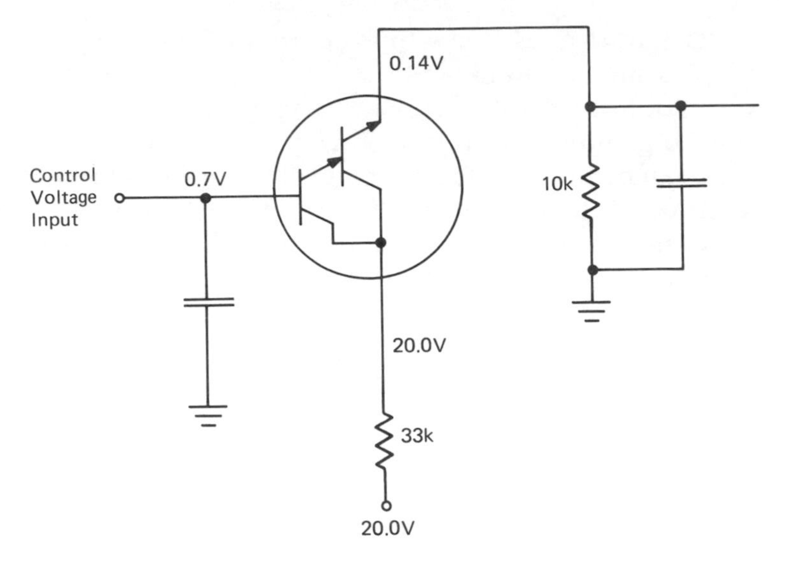

In this example, the Darlington transistor is normally forward-biased 0.56 V. However, collector current does not flow because this forward bias divides between the two junctions—each junction is forward-biased only 0.28 V.

Note: The Darlington color-killer transistor is normally cut off when chroma signal is present. (The 0.14 volt on the emitter terminal backs up from the load circuit.)

Figure 11-7 Color-killer arrangement. Darlington transistors provide very high current gain.

arrangement used in a Philco-Ford color-TV receiver is shown in Figure 11-7. It normally cuts off the chroma bandpass amplifier when energized by the color burst from the killer phase detector.

Note in Figure 11-7 that the transistor is normally cut off when a chroma signal is present. In other words, zero voltage drop is measured across the 33-kilohm resistor, and the collector voltage is only a small fraction of a volt. However, if the Darlington transistor becomes defective, a substantial leakage current can flow in the emitter circuit although the base is zero-biased or reverse-biased. In turn, an appreciable voltage drop will be measured across the 300-kilohm resistor with chroma signal present. The transistor then produces a weak-color or a no-color trouble symptom.

Refer to Figure 11-8. These are representative color-killer transistor terminal voltages in a Motorola color-TV receiver without

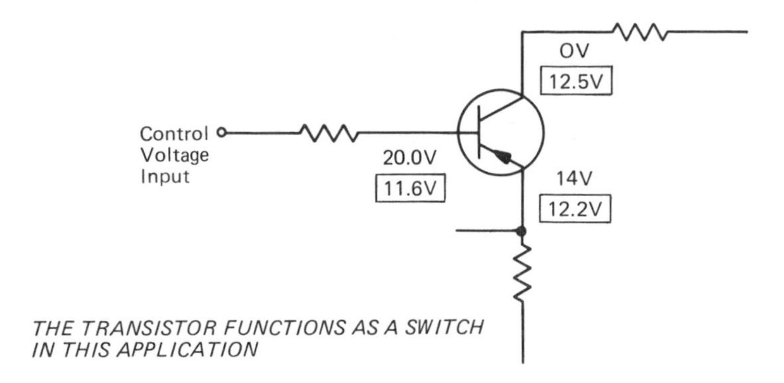

Device or component defects in the color-killer section cause symptoms of no-color reproduction, color drifting in and out, or weak-color reproduction.

Note: Representative normal voltage shifts on color-killer transistor from no-signal (vacant-channel setting) to keyed-rainbow input signal. The boxed voltages are normal for keyed-rainbow signal input. The unboxed voltages are normal for vacant-channel setting.

Figure 11-8 DC voltages change because of transistor nonlinearity. Example of color-killer transistor voltages with signal and without signal present.

---

signal present, and with keyed-rainbow signal present. Note that with chroma and black-and-white signal absent, the color-killer transistor is reverse-biased by 6 volts and is in collector cut off. In turn, the collector voltage is zero. On the other hand, with keyed-rainbow signal present, the transistor is forward-biased by 0.6 volt, and the collector voltage normally rises to 12.5 volts.

Although it might be supposed that the collector voltage would rise to V_{CC} when the transistor is cut off in Figure 11-8, it actually falls to zero because the transistor is operated in a grounded-collector circuit. This is just another way of saying that when V_{CC} is applied between the emitter and ground, the collector must go to zero potential when the base-emitter junction is reverse-biased. However, if the transistor is leaky, the troubleshooter will then measure more or less voltage at the collector terminal in the no-signal condition.

Bandpass Amplifier Troubleshooting

A bandpass-amplifier configuration operates as implied by its name, and passes the chroma-frequency band from 3.1 to 4.1 MHz.

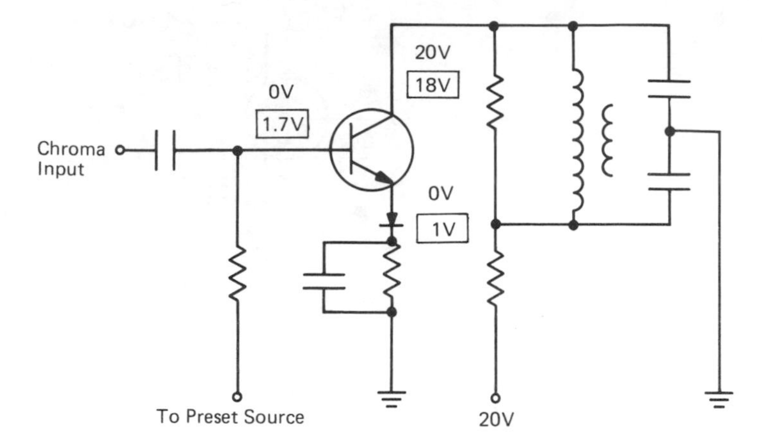

Note: Typical normal voltage shifts on bandpass-amplifier transistor from no-signal (vacant-channel setting) to keyed-rainbow signal input. The unboxed voltages are normal for vacant-channel setting.

Device or component defects in the bandpass-amplifier section result in weak- or no-color reproduction.

Figure 11-9 Example of bandpass-amplifier transistor voltages with signal and without signal present. Bandpass amplifier can be identified with a resonance probe.

Note in Figure 11-9 that the bandpass-amplifier transistor is normally zero-biased when the receiver is set to a vacant channel (no signal present). On the other hand, the transistor is forward-biased by 0.7 volt when a keyed-rainbow signal is present, and the collector voltage decreases by 2 volts.

Thus, the bandpass-amplifier transistor conducts zero collector current in the absence of signal input, but operates in class A with signal present. This example is from a Motorola color-TV receiver. DC voltage measurements provide preliminary troubleshooting clues, in case of bandpass-amplifier malfunction. Off-value DC voltages can result from either component or device defects, or from a fault in the preset base-bias source.

Accordingly, follow-up tests are often necessary following checkout of DC voltages. For example, note that the transistor can be checked in-circuit by means of a shut-off test. In other words, if the base and emitter terminals are temporarily short-circuited, the collector voltage will normally rise to 20 volts. If the collector voltage remains subnormal, the transistor is probably defective—but there is still the alternate possibility of leakage in a collector load capacitor. Therefore, the capacitors should be checked out before the transistor is unsoldered from its circuit.

Color Sync Gate and Amplifier

A color sync gate and amplifier is pulsed; it operates only during the color-burst interval. Thereby, the color burst is separated from the complete color signal and amplified for application to the subcarrier oscillator section. Defects in the color sync gate and amplifier section result in "touchy" color-sync action, or complete loss of color sync. When color sync is lost, diagonal red, green, and blue stripes or bars are displayed on the picture-tube screen. When color sync is marginal, false hues drift into and out of one or more areas in the color picture.

Note in Figure 11-10 that the color sync gate transistor is normally zero-biased when the receiver is set to a vacant channel (no

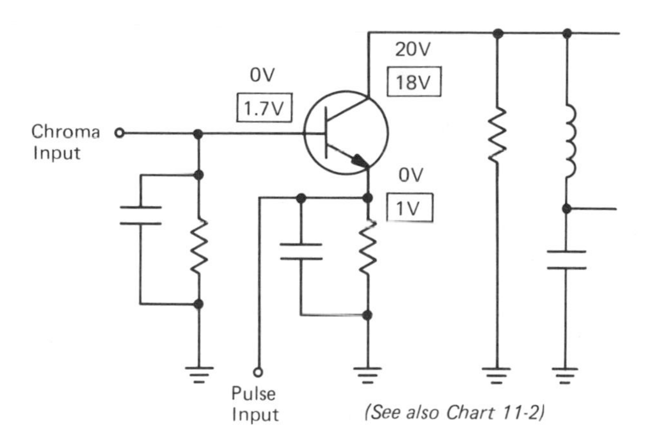

Note: Representative normal voltage shifts on the color sync gate and amplifier from no-signal (vacant-channel setting) to keyed-rainbow signal input. The unboxed voltages are normal for vacant-channel setting.

Color sync gate action can be checked to best advantage with an oscilloscope. (The pulse input waveform is sometimes off.)

Figure 11-10 Example of color sync gate and amplifier transistor voltages with signal and without signal present.

signal present). On the other hand, when a keyed-rainbow signal is present, the transistor is forward-biased by 0.7 volt, and the collector voltage decreases by 2 volts. The color sync gate transistor conducts zero collector current in the absence of signal, and the collector potential is at V_{CC}.

When chroma signal (keyed-rainbow signal) is present, the color sync gate transistor operates in class A. This is another example of chroma circuitry from a Motorola color-TV receiver. DC voltage measurements provide initial troubleshooting data, which often need to be supplemented by additional tests and measurements. For example, if all of the transistor terminal voltages are incorrect, the next step is to determine whether the transistor is defective, or whether there is a defective component in a branch circuit. Therefore, the troubleshooter should make a shut-off test of the transistor at this time.

If the transistor passes a shut-off test, it is cleared of suspicion. On the other hand, if the transistor fails a shut-off test, it is not necessarily defective—a leaky capacitor in the collector circuit can simulate collector junction leakage. Therefore, this possibility should be checked out before the transistor is replaced.

Color Demodulator Troubleshooting

Component or device defects in the color-demodulator section result in various trouble symptoms, as follows:

1. One color missing.
2. One color excessively intense.
3. One color weak (tint only).
4. All color reproduction absent.
5. One or two colors with incorrect hues.
6. Fluctuating or intermittent colors.

Note in the example of Figure 11-11 that all input terminals on the color demodulator module normally have the same DC voltage (4.5 V) in the absence of signal. Similarly, all input terminals normally have the same DC voltage (4.0 V) with keyed-rainbow signal present. The same basic relation is noted on the output terminals. All output terminals normally have the same DC voltage (8.4 V) in the absence

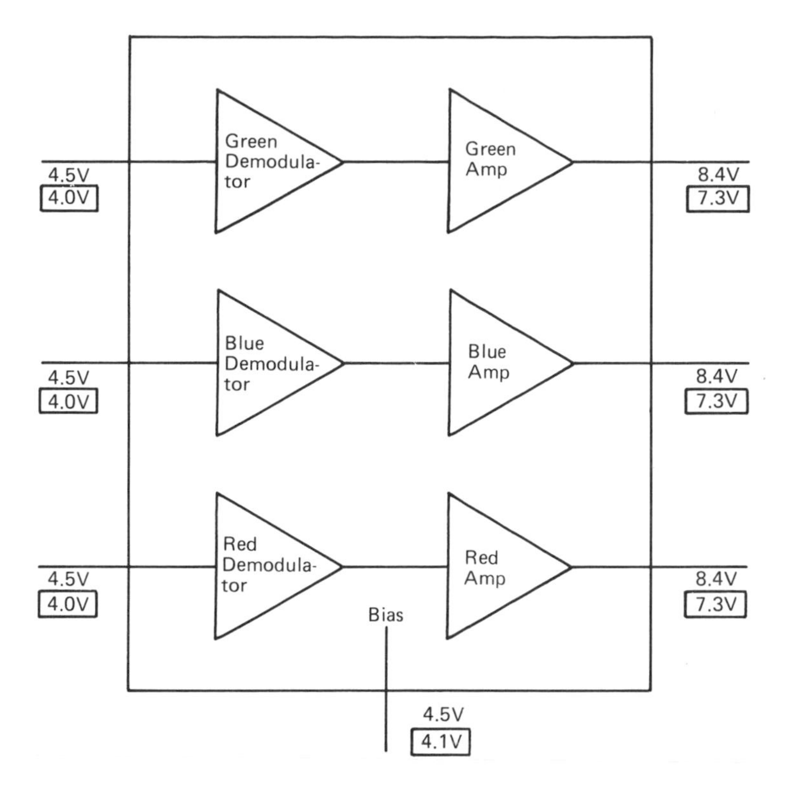

Note: Typical normal voltage shifts on the color-demodulator module from no-signal (vacant channel setting) to keyed-rainbow signal input. The unboxed voltages are normal for vacant-channel setting.

Color-demodulator circuit action is checked to best advantage with an oscilloscope. (See also Chart 11-2.)

Figure 11-11 Example of color demodulator module voltages with signal and without signal present.

of signal, and shift to 7.3 V with keyed-rainbow signal present. The demodulator bias line normally shifts from 4.5 V to 4.1 V.

When an input voltage incorrectly shifts value from the no-signal to the signal condition, it is probable that the trouble will be found in a preceding chroma signal section. On the other hand, when an output voltage incorrectly shifts value from the no-signal to the signal condition, the trouble will be found either within the module, or in the associated load circuit. *Preliminary troubleshooting generally consists in substitution of a known good module, to determine whether normal operation is resumed.*

If normal color reproduction resumes when the module is replaced, the defective module is either discarded or shipped to a repair depot. Some module failures are catastrophic, with the result that it is not economical to repair the damaged devices and components. On the other hand, some module failures are comparatively minor—for example, a single integrated circuit may have a defect. In such a case, it is more profitable to repair the module than to discard it.

12

Tape Recorder Troubleshooting

*Basic overview * Bias Voltage Consideration * Head Checkouts * Bias Voltage Checkout * Related Factors * Amplifier Troubleshooting * Variational Check * Frequency Distortion (Case History) * Hum Symptoms * Audio System Interference*

BASIC OVERVIEW

A *tape recorder,* in the strict sense of the term, has both recording and playback functions. A *tape player* has a playback function only. A *tape deck* denotes any tape unit that has no power amplifier or speaker; it usually has no housing and is intended for custom installation in a high-fidelity system. A *record/play deck* records tapes for playback through a hi-fi system, in a car, or through a portable tape player. A *play deck* is designed to play car tapes or for "second systems."

Although the troubleshooter is usually concerned with cassette tape units, reel-to-reel machines are encountered on occasion. Basic troubleshooting procedures are essentially the same for both types of machines. A block diagram is shown in Figure 12-1, with principal features noted. Careful distinction should be made between electronic malfunctions and mechanical problems in preliminary troubleshooting procedures. For example:

1. Weak or zero output on all tracks can be caused by a faulty tape; or, the tape head may be defective or out of adjustment; the head should be inspected for oxide deposits. (See Chart 12-1.)

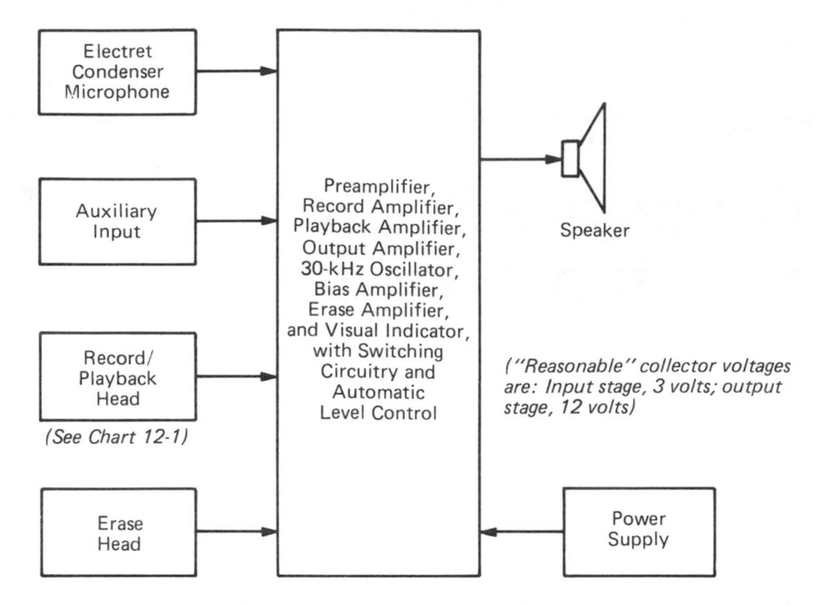

When troubleshooting without service data, it is helpful to recognize the standard configurations and to visualize their functional sections. This enables rapid selection of meaningful quick checks with identification of the relevant test points.

Although the troubleshooter may not have a full understanding of the circuit sections and functions of a particular machine, its stereo design permits effective "easter egging" by comparative voltage, resistance, and impedance checks.

Note: Smaller designs of cassette tape recorders employ simplified circuitry in which the playback amplifier also functions as the record amplifier. Similarly, smaller designs utilize a combination record/playback head, whereas separate heads may be provided in highly sophisticated designs.

Figure12-1 Block diagram for a cassette tape recorder. Nearly all tape machines are stereo designs.

2. Poor high-frequency response may be caused by an excessively worn head; the head should be inspected for oxide deposits; head alignment with respect to the tape should be checked.
3. Distorted audio output may be caused by a faulty tape; the speaker may be damaged or defective; batteries may be weak.

   In the absence of a mechanical problem, these three problems will be localized in the electronic circuitry.

CHART 12-1

Head Checkouts

A "reasonable" output voltage from a playback head in normal operation is 8 mV r.m.s.

Tape heads are prominent landmarks when troubleshooting without service data. Various trouble symptoms may be caused by defective tape heads.

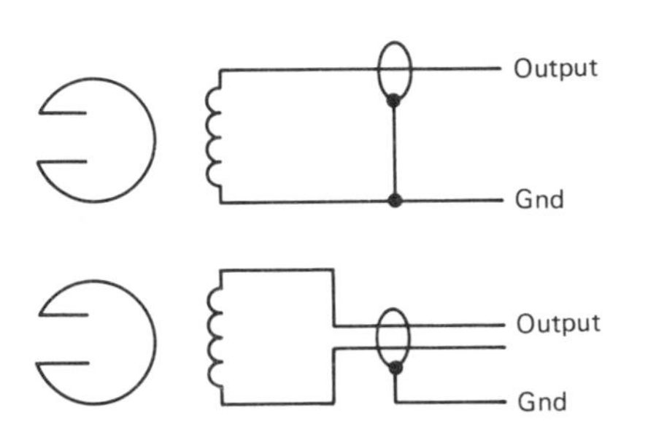

This type of tape head has a shielded connecting lead with a single inner conductor. (An erase head has unshielded leads.)

This type of tape head has a shielded connecting lead with two inner conductors. (An erase head has unshielded leads.)

Resistance checkout: The winding resistance of a typical tape head has a normal value of 514 ohms, with its output lead unsoldered from the preamplifier input. (An erase head normally has less resistance than a playback head, a record head, or a record/playback head—a typical erase head has approximately 1/3 the resistance of a playback head.)

Since nearly all tape machines are stereo designs, the troubleshooter can easily make comparative resistance measurements. In other words, it is highly improbable that both the L head and the R head will become defective at the same time. (Note that the leads to the L and R windings are color-coded.)

Impedance checkout: A follow-up impedance check of L and R head windings can be made as shown below:

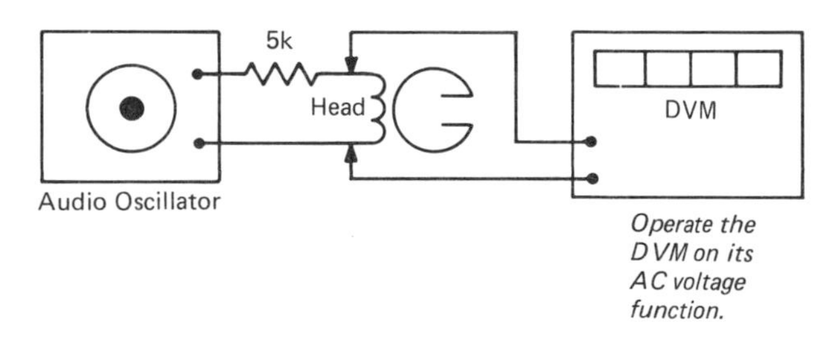

Operate the DVM on its AC voltage function.

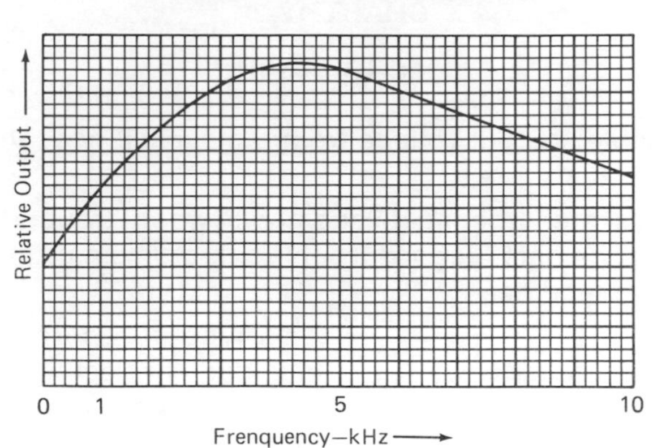

Impedance checks are made on a comparative basis. If DVM readings are taken at various frequencies, the head under test will normally show a resonant frequency (4500 Hz in this example). Note that if the test voltage is then increased, the resonant frequency will be higher (due to partial magnetic saturation).

4. Cross-talk between tape tracks is always caused by incorrect head adjustment, by a faulty tape, or by foreign substances in the cartridge opening.
5. Noisy output is likely to be caused by a magnetized playback head or by a faulty tape. An erratic tape transport can introduce noise. In the absence of a mechanical problem, however, noisy output will be localized in the electronic circuitry. For example, a transistor or resistor in the preamplifier may generate noise.

BIAS VOLTAGE CONSIDERATIONS

Since the oxide coating on a tape has a nonlinear magnetic characteristic, its transfer characteristic must be linearized by means of an AC bias voltage mixed with the audio signal. This AC bias voltage has a frequency in the 30 to 60 kHz range. Troubleshooting of the bias system involves the following considerations:

Optimum Amplitude: The bias voltage should be adjusted to optimum amplitude, although the bias-oscillator frequency is not critical.

Frequency Response: Note that the bias-voltage amplitude affects the frequency response of the transfer characteristic, as well as its linearity. (See Chart 12-2.)

Output Level: The bias-voltage amplitude also affects the output level of the recorded tape on playback.

Peak Bias: The peak-bias adjustment corresponds to a bias-voltage amplitude that provides maximum output level on playback.

Overbias: Maximum linearization (minimum harmonic distortion) is obtained by overbias—approximately 2 dB over peak bias.

Low-frequency Response: Optimum low-frequency response is obtained at peak-bias amplitude.

High-frequency Response: High audio-frequency response is reduced by overbias.

Underbias Effects: Underbias adjustment results in distorted output, low signal-to-noise ratio, and reduced output level.

Thus, the AC bias-voltage adjustment involves certain conflicting factors, and the optimum value is necessarily a compromise.

RELATED FACTORS

Elaborate cassette decks often provide a front-panel adjustable bias control, whereby the operator can optimize the sound output for conventional, chrome, or metal tape. The maximum bias-voltage level is typically 0.7 r.m.s. volt. Note that poor high-frequency response can be caused by a reversed tape, as well as by an excessively high bias voltage. A magnetized recording head, defective microphone, or even an incorrect setting of the tone control can also cause poor high-frequency response.

Although less common than mechanical or amplifier defects, a distorted bias waveform can also cause noisy and/or distorted sound

CHART 12-2

Bias Voltage Checkout

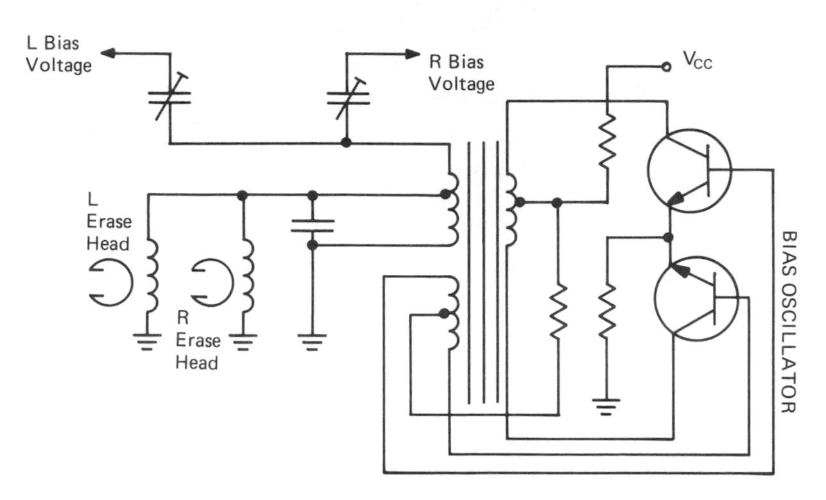

The bias and erase voltages are operative only in the record mode. The bias voltage is adjustable in all but the simplest tape machines. A maximum bias voltage of 700 millivolts is typically available.

An erase voltage of 15 volts r.m.s. is typical, at a frequency of 60 kHz. An oscilloscope with calibrated time base can be used to check the frequency of the bias voltage, and also its waveform. The bias voltage should have an approximately sinusoidal wave shape, for optimum fidelity and signal-to noise ratio. If the bias waveform is substantially distorted, check the bias oscillator circuitry.

Note that the simplest tape machines may employ permanent-magnet or electromagnet erase facilities.

Note in passing that the more elaborate tape machines employ separate recording and playback heads. Less elaborate machines utilize a compromise head that serves for both record and playback operation.

A compromise head has a gap width intermediate between that of separate recording and playback heads. Although the simplified design serves adequately for utility applications, it does not provide high-fidelity operation.

To determine whether a head is excessively worn, make a

CHART 12-2 CONTINUED

comparison check of the gap under a magnifying glass with respect to the gap in a new head. In this area of troubleshooting, experience is the best guide.

When a new head is installed, make certain that the azimuth/height adjustments are correct. An incorrect height adjustment permits adjacent-channel cross-talk. An incorrect azimuth adjustment causes loss of high audio frequencies.

reproduction. Waveform distortion results from a faulty device or component in the bias-oscillator section.

Hum interference in the output points to defective ground connections—check the braid connection on the microphone cable. In line-operated equipment, the power-supply filter section is a prime suspect. Some decks are provided with polarized power plugs—if the broad prong is "trimmed" or otherwise defeated, audible hum may result from reversed insertion of the plug into a power outlet.

AMPLIFIER TROUBLESHOOTING

Basic audio troubleshooting techniques were explained in Chapters 1-3. Note that tape-recorder and high-fidelity amplifiers may employ negative-feedback loops—in some situations, a feedback loop can be a tricky act to follow. Consider, for example, the arrangement depicted in Figure 12-2. As previously noted, distortion shows up in a sine-wave test as an abnormal or subnormal peak/average voltage ratio. In the example, the worst peak/average voltage ratio will be measured at the preamplifier although the driver stage is actually defective.

In turn, unless the feedback loop is kept in mind, the troubleshooter will falsely conclude that the preamplifier stage is defective, because its output waveform is badly distorted, compared with the driver output waveform.

Analysis of the configuration in Figure 12-2 reveals that the 2-percent distortion from the output section produces 20-percent distortion when mixed with the low-level input waveform. Essentially, the 20-dB negative-feedback loop predistorts the input waveform to the preamplifier. This distortion cancellation is incomplete—

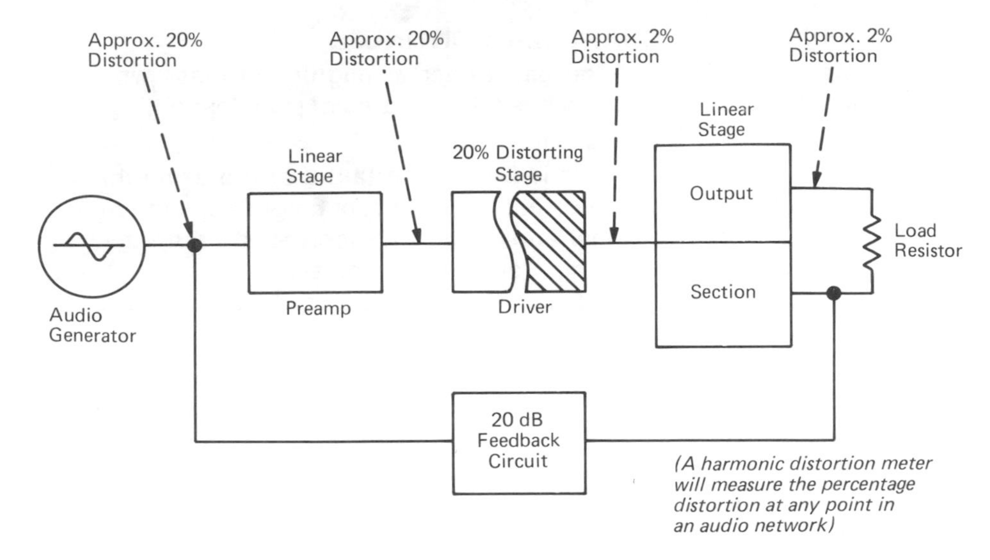

Percentage distortion is generally measured with a harmonic-distortion meter. It is expressed as total harmonic distortion (THD). Percentage distortion can also be measured with an intermodulation analyzer (two-tone distortion checker). The measurement is expressed as percentage intermodulation distortion (IM). Harmonic distortion and intermodulation distortion percentages are approximately the same.

Note: 20 dB of negative feedback denotes that the feedback voltage reduces the input signal voltage by 20 dB from its value with the feedback loop open-circuited. In other words, 20 dB of negative feedback reduces the input signal voltage to 0.1 of its initial value. Therefore, the feedback amplifier must provide 10 times as much gain as would be required for the same output level without negative feedback.

Figure 12-2 Although the driver stage is defective, the preamp looks like the defective stage.

2-percent distortion remains in this example, which in turn appears at the system output.

VARIATIONAL CHECK

A variational check for audio stage distortion is shown in Figure 12-3. An audio generator is connected to the input terminals, and a DC voltmeter is connected at the output terminals of the stage. To check for distortion, the generator output is increased from zero to the maximum rated input level for the stage. As the signal level is

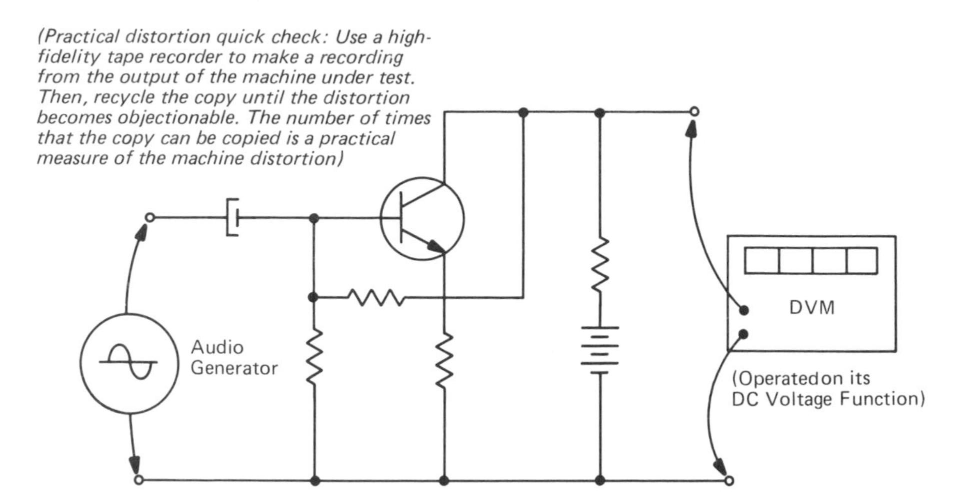

If the stage is operating in class A at maximum rated power output, there will be no shift in DC level from the minimum to the maximum drive condition. A shift in DC level indicates that the transistor is operating in part as a rectifier (nonlinearly). As a cautionary note, this is not a completely informative test. If the transistor happens to compress or to clip both the positive peaks and the negative peaks of the sine wave by the same amount, there will be no shift in DC level, although distortion is taking place.

As another cautionary note, an oscilloscope has limited ability to provide adequate indication of small percentages of distortion. For example, it is difficult to see 2-percent distortion in a displayed sine wave.

Note: This test setup is also used to check the frequency response of an audio amplifier. However, the DVM is then operated on its AC voltage function, with a series blocking capacitor.

Figure 12-3 Test setup for variational analysis of stage distortion.

varied, the DC voltmeter reading is observed. *If any shift occurs in the voltmeter reading, the troubleshooter concludes that the stage is operating in a nonlinear manner.*

The principle of this variational check for distortion is based on the partial rectification that occurs when a signal passes through a stage that has a nonlinear transfer characteristic. Compression of the peak signal excursion results in more or less rectified signal current flow through the collector load resistor. In turn, the average DC in the collector circuit changes, and a shift in collector voltage occurs.

Note that a variational check is not conclusive proof that a stage is free from distortion. In other words, if a shift in collector voltage occurs, the troubleshooter knows that the stage is distorting. On the other hand, if no shift in collector voltage occurs, the stage may nevertheless distort the signal. If both the positive peak and the negative peak of the signal happen to be equally compressed, no shift in DC collector voltage will occur, although the signal is being distorted.

FREQUENCY DISTORTION (CASE HISTORY)

A nonlinear transfer characteristic causes amplitude (harmonic) distortion. Frequency distortion, on the other hand, is generally caused by defective capacitors. For example, consider the case history depicted in Figure 12-4. The sound output had a "tinny" quality and lacked lower audio frequencies. DC voltage values were normal, and a low-power ohms check showed that the resistive values were within tolerance. A bias-on test indicated that the transistor was workable.

Then, it was recognized that the 0.0015μF capacitor in the negative-feedback circuit from collector to base functions as a frequency-responsive component (provides increasing negative feedback at higher audio frequencies). Accordingly, the troubleshooter checked the feedback capacitor and found that it was open. Replacement of the capacitor restored the amplifier to normal frequency response.

Hum Symptoms

Low-level circuits are much more susceptible to hum and interference pickup than are high-level circuits. An audio system can be divided in a general way into low-level circuits and high-level circuits. For example, the output circuit from a tape recorder is a low-level circuit. On the other hand, the output circuit from a preamplifier is a comparatively high-level circuit.

Most of the interconnections in an audio system are made with audio cable. This is a shielded and stranded type of conductor; the shield braid that surrounds the central conductor is grounded to the chassis (or to the common supply bus) of the audio equipment. Thereby, undesirable pickup of hum and various other stray fields is virtually eliminated.

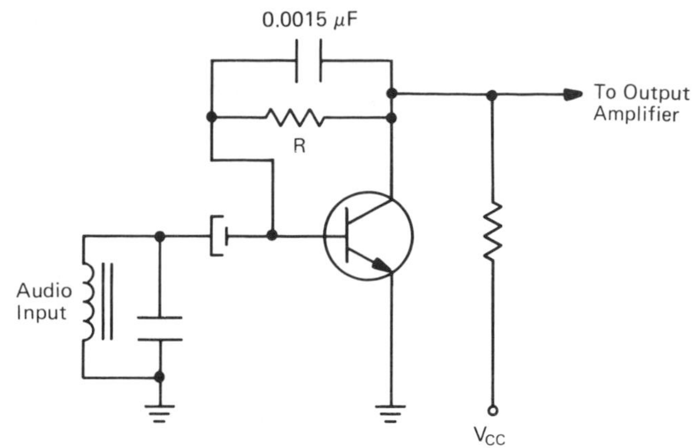

Note: Frequency distortion occurs when the 0.0015μF feedback capacitor is open.

This is an example of a frequency-selective negative-feedback arrangement. In other words, the reactance of the feedback capacitor decreases as the frequency increases, thereby developing less stage gain at higher frequencies.

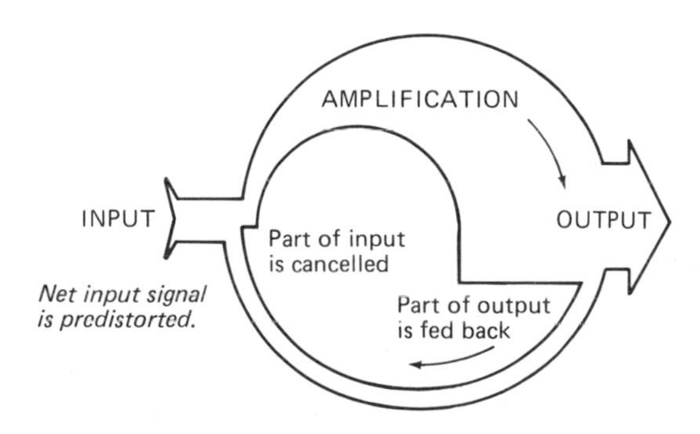

Note that resistor R also provides negative feedback from collector to base. This negative feedback component, however, is not frequency-selective, inasmuch as the resistor has the same value at all audio frequencies.

Figure 12-4 Example of a frequency-compensated stage. Case history of distortion caused by an open capacitor.

Occasionally, in spite of observance of all good practices in installation procedures, an audio system may pick up objectionable interference. This trouble symptom results from operation in a location that has abnormally high stray-field intensity. In turn, the low-level circuits must be very well shielded, and the braid of the associated audio cables must be well grounded. In addition, it is good practice to connect a separate ground wire from a record player to its preamplifier. This should be a heavy copper wire to provide a low-impedance ground return for the phono circuit, and thereby minimize the possibility of audible hum from this source.

AUDIO SYSTEM INTERFERENCE

Audio system interference, also called RFI, is the result of "audio rectification." This interference consists of radio-frequency energy pickup by an audio amplifier, or by a component connected to the amplifier. In a high-level radio-frequency field, an audio system can process sufficient induced signal voltage that distorted sound output and audio interference occurs. *Audio rectification denotes unexpected demodulator (detector) action in a normally linear amplifier.*

Audio rectification is the result of overdrive (usually in the input stage). When overdrive occurs, a transistor or an integrated circuit can produce sufficient audio rectification that an objectionable amount of interference is passed through to the remainder of the audio system and to the speaker. Correction of audio rectification usually consists of bypassing, shielding, filtering, and/or choking arrangements added into the amplifier circuitry, as in the example of Figure 12-5.

> ***An RF choke used for suppression of RFI may have a value from 5 to 7 microhenries*** **(μH) in the range from 30 to 110 MHz, or 1.5 μH in the range from 80 to 200 MHz. A 250-picofarad (pF) bypass capacitor is generally suitable for RFI suppression.**

In most cases, audio rectification occurs in the first stage of a preamplifier, because the input stage operates at the lowest level and is most susceptible to overloading. The input stage is also followed by a high-gain amplifier system, so that any interference developed by the first stage is stepped up greatly in the following portion of the system.

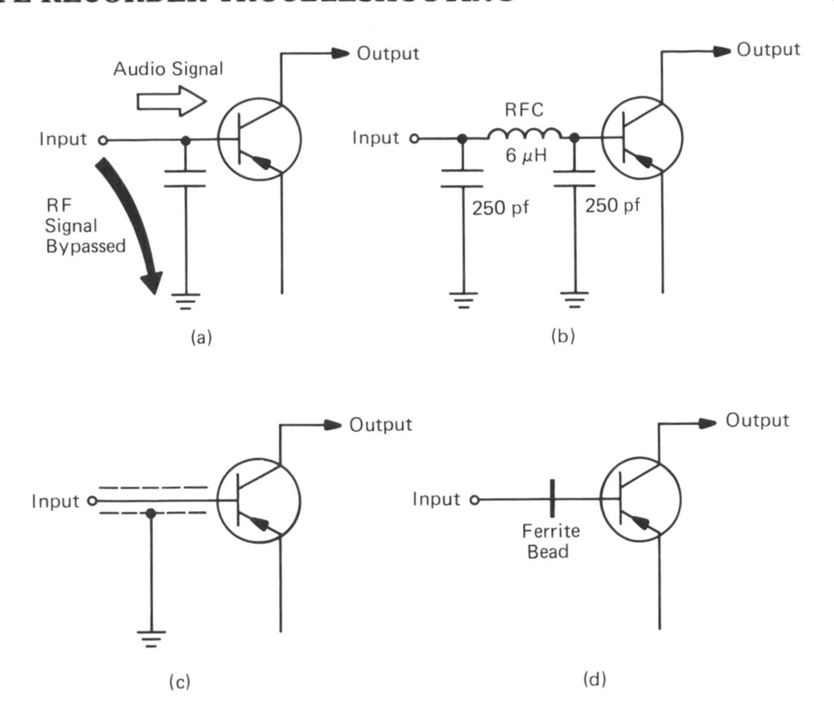

RFI may be intense in the near-field of a broadcast station.

Figure 12-5 Basic RFI filtering arrangements. (a) Bypass capacitor in base circuit; (b) pi-filter section in base circuit; (c) shielded input lead in base circuit; (d) ferrite bead choke in base circuit.

When RFI is encountered, it is often helpful to check its level as the volume-control setting is varied. If the interference is minimized when the volume control is turned to its minimum position, the troubleshooter concludes that the point of entry is located ahead of the volume control. On the other hand, if the volume-control setting has no effect on the interference level, the troubleshooter concludes that the point of entry is in the circuitry following the volume control. In some situations, it will be found that part or all of the RFI interference is entering via the power cord to the outlet. In this situation, a pi-type power-line filter, as shown in Figure 12-6, may be effective. Note that the 50-mH choke coils must carry the total current demand of the power supplies in the audio system; in turn, they must be suitably rated for the application.

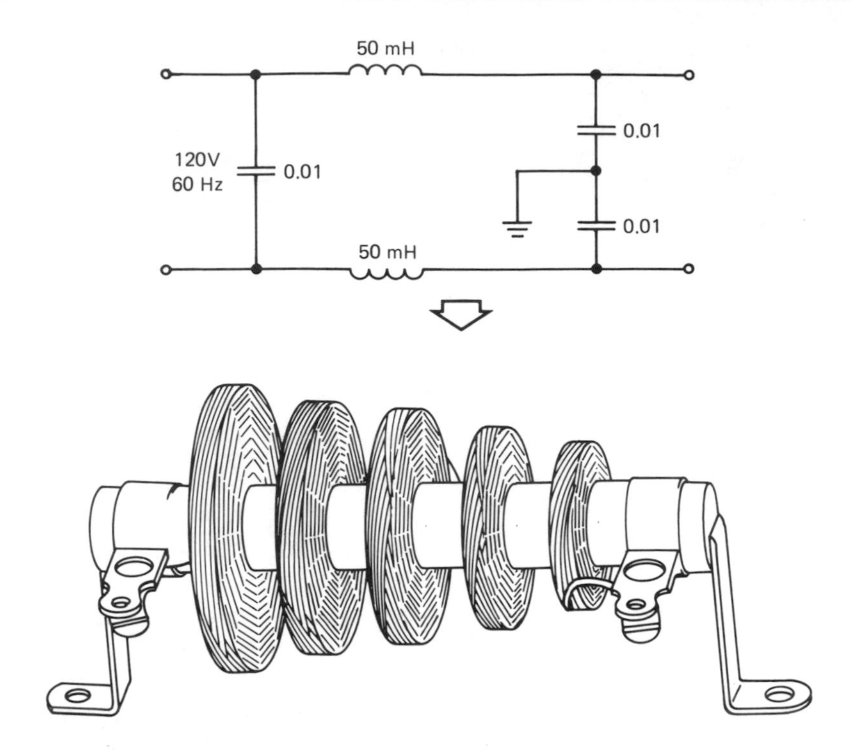

Metropolitan power lines tend to have high RFI levels.

Figure 12-6 Pi-type line-filter arrangement.

13

Intercom and Telephonic Equipment Troubleshooting

*Preliminary Trouble Analysis * Wired System * Quick Checks * DC Voltage Measurements * In-Circuit Resistance Measurements * Examples of Ohmmeter Crawl * Input Sensitivity Measurement * FM Wireless Intercom * Telephone Listener * Self-Oscillatory Quick Check*

PRELIMINARY TROUBLE ANALYSIS

An intercommunication system is basically a two-way telephonic arrangement with a microphone and (usually) a loudspeaker at each station. The loudspeaker commonly serves also as a microphone. Wired intercom systems employ cables or lines for station interconnection. Wireless intercom systems utilize carrier-current communication on power wiring between stations. Carrier currents have frequencies in the ultrasonic range; either amplitude modulation or frequency modulation may be employed. (See Chart 13-1.)

Wired System

Wired intercom systems may be battery operated or line operated. Preliminary trouble analysis proceeds as follows:

1. When feasible, discuss the onset and nature of the trouble symptom(s) with the intercom owner or operator.
2. Check the battery voltage (if intercom is battery operated).
3. Check the line continuity between stations.
4. Observe the responses to operation of the intercom controls.

CHART 13-1

Preliminary Troubleshooting without Service Data

Troubleshooting without service data can be accomplished in various situations by observing the following steps:

1. Consider the basic function(s) of the unit—whether it is self-contained or whether it operates in combination with other units; whether it is a wired design or a wireless design; whether it is battery powered or line powered.
2. Evaluate the trouble symptom(s) and make an educated guess concerning the malfunction.
3. Make click tests on suspected transistors—apply a temporary short-circuit between the base and emitter terminals of the transistor.* Listen for a click from the speaker.
4. If a transistor fails a click test, measure its collector voltage, and measure its base-emitter bias voltage to determine whether the readings are reasonable.

Transistor fails click test, but its terminal voltages are reasonable: The fault is most likely to be in an associated branch circuit—for example, a coupling capacitor may be open.

Transistor fails click test, and its terminal voltages are not in the ballpark: The fault may be either in the transistor or in an associated branch circuit—for example, the transistor may have a shorted or open junction; a capacitor in a branch circuit might be defective, or a resistor in a branch circuit might be burned.

Transistor passes click test, but its terminal voltages are not in the ballpark: The fault is most likely to be in an associated branch circuit—for example, a defect in the base circuit or in the emitter circuit can put the transistor into deep saturation, but it will still pass a click test.

*Refer back to Chart 1-1.

5. Open the case or cabinet and try to locate the fault by sight, sound, overheating, or odors.
6. If the chassis is dusty or dirty, clean it thoroughly—in various cases, the trouble will be spotted before the cleaning job is completed.

QUICK CHECKS

If further trouble analysis is necessary, useful quick checks can be made, as follows:

1. *Click Tests.* Hold the speaker of the master unit close to your ear, and turn the power switch off and on. A click is normally audible. With the power switch turned on, press the "talk" button—a click is normally audible.
   Repeat the click test for the substation or remote unit. These tests can help to localize the general trouble area.
2. *Current Demand.* Disconnect one of the battery clips and insert a milliammeter in series with the battery circuit. There is normally almost zero current demand when the power switch is turned off. *If current does flow, look for leaky electrolytic capacitors.* (A typical 2-station intercom normally draws only 1.5μA from the battery when the power switch is turned off.)
3. Next, turn the power switch on and check the operating current demand. A typical 2-station intercom drew 6mA when tested, although a drain as high as 15 mA would have been within rated tolerance, as noted in the service manual. A current demand of 3 mA, or of 20 mA, would be regarded as a trouble symptom.
   Excessive current demand can result from accidental reversal of battery polarity. In turn, electrolytic capacitors may be damaged, and transistors may fail catastrophically. Then, when the battery is reconnected in correct polarity, excessive current demand may persist.
4. If the current demand is within rated tolerance, proceed to turn up the volume control. Hold the speaker close to your ear and listen for noise output. If there is no audible noise, it is indicated that there is a dead or weak audio stage in the amplifier system.
5. A supplementary test can be made by turning up the volume control and bringing the substation unit near the master unit. If the amplifier system is developing appreciable gain, acoustic feedback will occur, and the speaker will "sing" or "howl." Any audible output, no matter how weak, noisy, or

distorted, can give helpful clues concerning the trouble location.

Case History

With the power switch turned off, a weak motorboating sound was audible from the master speaker. Current demand (6 mA) was normal, provided that the power switch was on—but the power switch was off. The rated current demand with power switch off was only 1.5μA. Although suspicion fell on the power switch, it was not defective. The fault was finally tracked down to an electrolytic capacitor in the substation unit. This capacitor had a leakage resistance of 3,000 ohms. (See Figure 13-1 and Chart 13-2.)

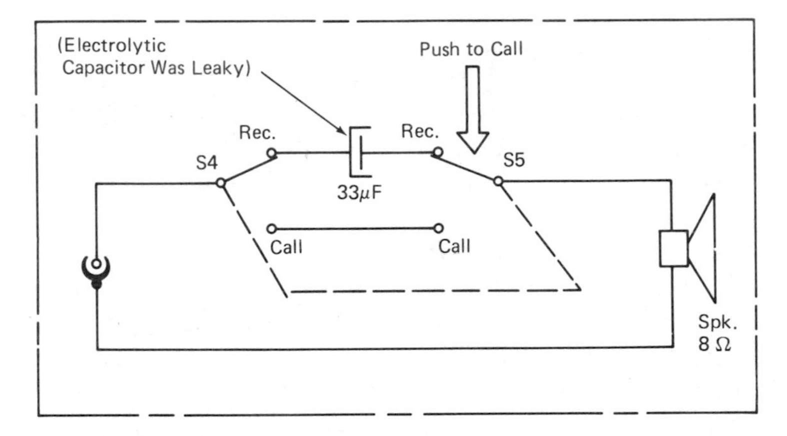

Note that leaky devices or components can sometimes be picked up on the basis of comparative temperature tests, as explained in Chapter 1.

Open capacitors, or capacitors with a poor power factor can be spotted with an impedance checker, as explained in Chapter 2.

(A suspected open capacitor can be quick-checked by "bridging" it with a known good capacitor of the same type. If normal operation resumes, the suspicion is confirmed.)

If a unit has been in service a long time, the switches should be tested for defects.

Figure 13-1 Typical intercom substation configuration. Two substations may be available for comparative tests.

CHART 13-2

Preliminary Resistance Checkouts

Helpful quick-check trouble clues can be picked up on occasion by means of a preliminary resistance checkout. *The most important test in the case of a battery intercom is the resistance between the battery-clip terminals.* Proceed as follows:

1. Disconnect the battery from the intercom unit.
2. Turn the power switch on.
3. Measure the resistance between the battery-clip terminals, with ohmmeter polarity the same as normal battery polarity. (Use high-power ohms function.)
4. Compare the measured resistance values on adjacent ohmmeter ranges, and on the various switch positions.

Example: A normally operating standard 4-transistor master intercom unit showed the following resistance values:

Power Swtich On: Rx10k range, 10 kilohms, Rx1k range, 13 kilohms. (No resistance change when "Call" or "Press to Talk" switches are thrown.)

Power Switch Off: Rx10k range, 10 megohms; Rx1k range, infinite. (Resistance reading changes to 22 kilohms on Rx10k range when a "Call" or "Press to Talk" switch is thrown. Resistance reading changes to 50 kilohms on Rx1k range when a "Call" or "Press to Talk" switch is thrown.)

Although there is a reasonable tolerance on the battery circuit resistance from one intercom to another, substantially higher or lower measured values would point to a defect in the battery load circuit.

Note that when the foregoing tests are made with a low-power ohmmeter, the battery circuit resistance normally measures infinity.

TROUBLESHOOTING WITHOUT SERVICE DATA

Success or failure in troubleshooting without service data is closely related to the troubleshooter's knowledge of circuit action, experience, and resourcefulness. An attitude of patience, curiosity, and involvement is also a key factor.

Note that troubleshooting without service data can be a breeze if a similar unit in good working condition is available. In such a case, the troubleshooter merely makes systematic voltage and resistance measurements through the two units, and compares their values. When measured values indicate a fault is present in a particular circuit, individual components and devices are checked out to pinpoint the failure.

DC VOLTAGE MEASUREMENTS

When preliminary tests do not turn up the cause for an intercom malfunction, the troubleshooter generally proceeds to make DC voltage measurements. With reference to Figure 13-2, note the following relations:

1. All four transistors are normally forward-biased.
2. All transistor voltages shift when the master and substation switches are operated.
3. The amplifier is essentially a current (or power) amplifier, rather than a voltage amplifier.
4. Input, driver, and output transistors are all direct-coupled.

DC voltage values should be checked for the switch positions noted in the diagram. Measured values normally agree with ±20 percent of the specified values, except that the tolerance is much tighter on the base-emitter bias voltages. The difference between a measured base voltage and a measured emitter voltage should agree very closely with the difference between the corresponding specified base and emitter voltages. Although the amplifier configuration is direct-coupled, and an error voltage in one circuit branch will be reflected to some extent into other circuit branches, the greatest upset the DC-voltage distribution will occur in the circuit associated with the defective device or component.

When DC voltage and resistance measurements do not suffice to close in on the defective device or component, an audio signal tracer like that shown in Figure 2-1 can often save the day. The signal tracer shows where the signal is stopped (or seriously attenuated).

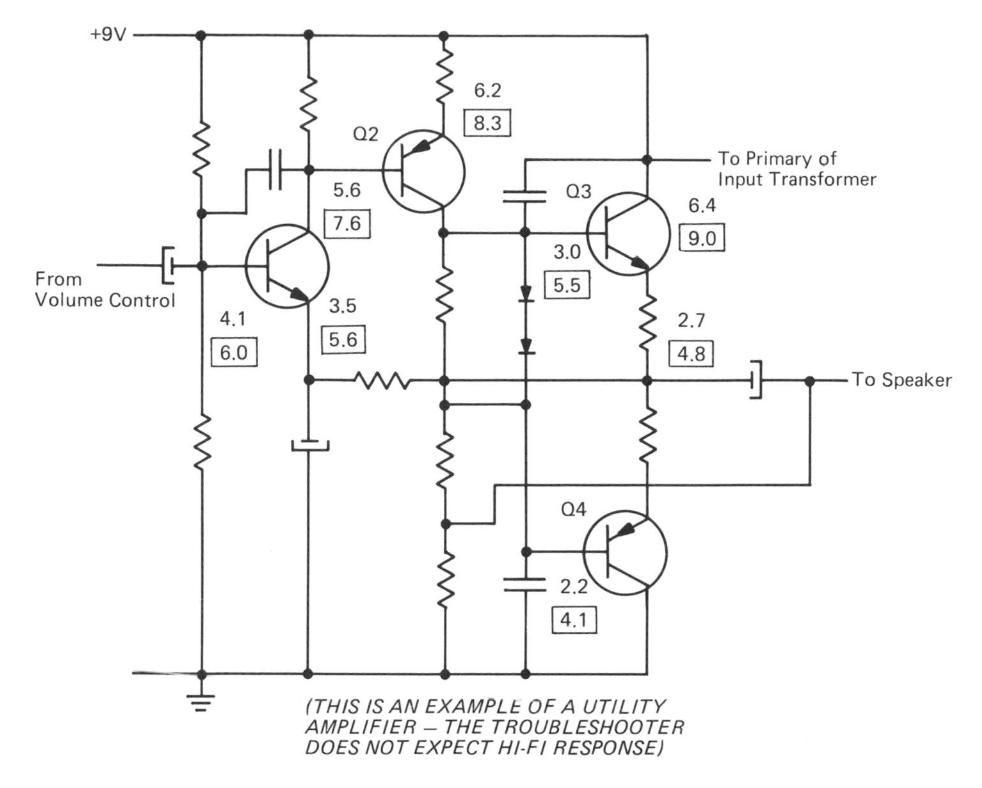

When troubleshooting without service data, buzz out the transistors as explained in Chart 1-1. Evaluate the voltages in terms of ball-park values, keeping in mind that class-A operation is used in this type of equipment.

Note: Boxed voltages are normal for master switch in "Call" position, and substation switch in "Receive" position. Unboxed voltages are normal for master switch in "Receive" position, substation switch in "Call" position.

Make comparative temperature checks, if a duplicate master is available.

Figure 13-2 Amplifier configuration for a typical 4-transistor intercom master unit.

Alternatively, the troubleshooter can work back and inject a test signal at the bases of various transistors, using the output from an audio oscillator.

IN-CIRCUIT RESISTANCE MEASUREMENTS

When DC-voltage measurements do not identify a circuit defect, the troubleshooter generally proceeds to make in-circuit resistance measurements with a low-power ohmmeter. In a configuration like that shown in Figure 13-2, a resistance may not be immediately indicated by an ohmmeter—the reading may "crawl" as an electrolytic capacitor gradually charges through a resistor. This is a normal test condition—if the reading did *not* crawl, it would be indicated that the capacitor is open.

Examples of Ohmmeter Crawl

1. The leakage resistance of a 100μF electrolytic capacitor is measured on the Rx10k range of a high-power ohmmeter. The ohmmeter crawls rapidly at first, and then slows down. At the end of 1 minute, the reading is 100 kilohms. At the end of 2 minutes, the reading is 150 kilohms.
2. The foregoing capacitor is then discharged, and its leakage resistance is remeasured. Now, the ohmmeter crawls much more rapidly. At the end of 1 minute, the reading is 175 kilohms. (The first measurement provided some forming action.)
3. When the leakage resistance of the capacitor is measured on the Rx1k range of the ohmmeter, the ohmmeter crawls still more rapidly. This speed-up of ohmmeter response results from the lower internal resistance of the instrument on its Rx1k range.
4. The capacitor is then discharged and connected in series with a 30-kilohm resistor. The resistance of the series circuit is then measured on the Rx1k range of the ohmmeter. As soon as the test leads are applied, the ohmmeter indicates 30 kilohms—the value of the series resistor. However, the ohmmeter then slowly crawls as it approaches the sum of the series resistance and the leakage resistance.
5. The capacitor is again discharged, and the resistance of the series circuit is remeasured using a low-power ohmmeter. As soon as the test leads are applied, the ohmmeter indicates 30 kilohms, and then slowly crawls to its final readout, as before. This is just another way of saying that in-circuit resistance measurements may require appreciable "settling time" in some cases.

INPUT SENSITIVITY MEASUREMENT

The input sensitivity of an intercom denotes the input signal voltage that is required to produce rated output. For example, the intercom amplifier depicted in Figure 13-2 is rated for an input sensitivity of 0.18 mV r.m.s. for an output of 50 mW (0.63 V r.m.s. across an 8-ohm load resistor, or speaker). The input sensitivity is checked as shown in Figure 13-3.

An audio generator followed by a 100-to-1 attenuator is used. The test is made at the standard frequency of 1 kHz. An 18 V r.m.s. output level from the generator is attenuated to 0.18 V r.m.s. at the intercom input. (The effective generator resistance is approximately

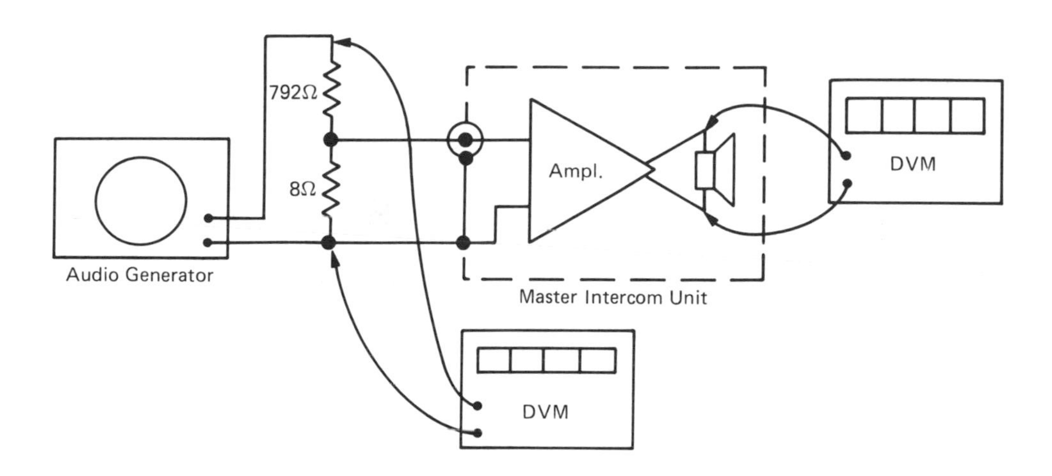

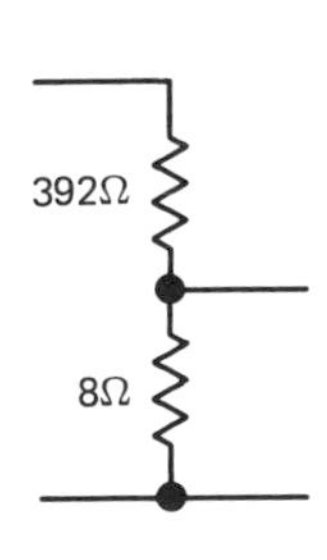

Note: If the audio generator cannot supply 18 volts, but can supply 9 volts, a 50-to-1 voltage divider may be used, consisting of a 392-ohm resistor and an 8-ohm resistor. In turn, a 9-volt signal applied to the attenuator is reduced to 0.18 volt for application to the intercom. Resistors used in an attenuator should have a tolerance of ±1 percent.

When troubleshooting without service data, and without a comparison unit, useful reference values (ball-park values) can often be obtained from somewhat similar intercom units. In other words, intercoms that do the same thing in the same general way can provide various comparable test data.

Note that an intercom has a typical frequency response of 200-3000 Hz.

Figure 13-3 Checking the sensitivity of the master intercom unit.

8 ohms in this arrangement.) If the intercom amplifier has rated sensitivity, 0.63 V r.m.s. will be measured across the speaker terminals. Note that in this example, the input sensitivity could be as low as 0.25 V r.m.s., and the amplifier would still be within tolerance.

This is a comparatively small voltage gain (3.5 times). However, it represents a *power gain* of 12.25 times, or more than 10 dB. Since the input and output resistances are the same in this example, the current gain is the same as the voltage gain (normally 3.5 times). (The power gain is equal to the product of the voltage gain and the current gain.) The rated power output from the master intercom is 50 mW ($0.63^2/8$).

Case History—Low Input Sensitivity

A wired intercom with the amplifier configuration shown in Figure 13-2 developed low input sensitivity. When the substation unit was brought near the master station unit, acoustic howl did not occur unless the volume control was turned to maximum. The battery voltage was normal. When the transistor terminal voltages were measured, it was found that the collector voltage on Q1 was very low, and read 0.67 volt. The troubleshooter suspected that Q1 had collector-junction leakage. When tested, the transistor showed a collector-base leakage resistance of 20 kilohms, approximately. Accordingly, Q1 was replaced and the intercom was in turn restored to normal operation.

FM WIRELESS INTERCOM

The arrangement of an FM wireless intercom is shown in Figure 13-4. When the unit is plugged into a 117-V line, the signal is coupled into or out of the line via a pair of 0.0047 μF coupling capacitors. The carrier signal has a frequency of 200 kHz, in this example, and is frequency modulated. *Frequency modulation is preferable to amplitude modulation, in that an FM wireless intercom is comparatively immune to line noise.* Note, however, that heavy switching surges on the line will normally be audible as clicks in the sound output.

As in most intercom configurations, the speaker also serves as a microphone. In the example of Figure 13-4, the incoming signal from the line is first stepped up through the 200-kHz amplifier, and is then demodulated through the locked oscillator-detector. In turn, the

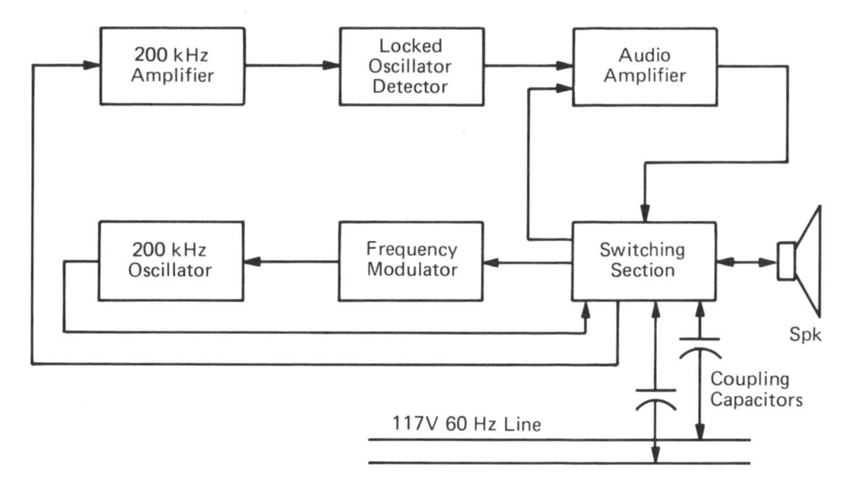

Although troubleshooting without service data is facilitated by an understanding of the standard configurations and sectional functions of FM wireless intercom units, effective "easter egging" can be accomplished by comparative measurements on identical units.

Note: A general troubleshooting rule states that a malfunction will usually be due to a single component or device defect. Although this is often true, exceptions will occur. For example, a shorted capacitor may produce an excessive current flow, with the result that an associated resistor is burned. Similarly, a short in an output transistor can result in overheating of associated resistors.

Figure 13-4 Arrangement of an FM wireless intercom unit. Identical units are almost always available.

audio signal is stepped up through the audio amplifier and is then applied to the speaker.

When the FM wireless intercom is switched to its "Talk" function, the speaker serves as a microphone, and its output is stepped up via the audio amplifier which now functions as a speech amplifier. In turn, the audio signal is applied to a varicap diode circuit which frequency-modulates the 200-kHz oscillator section. The output from the modulated oscillator, in turn, is coupled into the line.

"Dead Intercom"

A "dead" trouble symptom can denote no transmission, no reception, and no call tone action. In such a case, the troubleshooter should first check out the installation and make certain that there is no more than 200 feet of line between the units, and also that the line for each unit is connected to the same side of the power distribution transformer.

"No Transmission, Reception OK"

When the trouble symptom is "No transmission, reception OK," the malfunction is localized to the transmitter circuitry. The 200-kHz oscillator is a ready suspect—the troubleshooter should make a quick check for oscillator activity, as depicted in Figure 13-5. The 200-kHz oscillator has considerable harmonic output, with the result that when an AM radio receiver is placed near the intercom's line cord, heterodyne interference with station carriers is normally obtained as the tuning dial on the radio receiver is varied. *If no heterodyne interference occurs, the trouble will be found in the intercom oscillator section.*

"No Modulation, Reception OK"

When the trouble symptom is "No modulation, reception OK," the malfunction is most likely to be in the modulator circuit, although it might also be localized in the switching section. With reference to Figure 13-5, note the following test results:

1. In normal operation, when the AM radio receiver is tuned to a harmonic of the 200-kHz oscillator, the "Call" tone will be reproduced by the speaker in the radio when the intercom's "Call" button is depressed.
2. In normal operation, incidental amplitude modulation is present in the frequency-modulated output from the intercom. Thus, when the AM radio receiver is tuned to a harmonic of the 200-kHz oscillator, the sound of a fingernail moving across the speaker grille of the intercom will normally be reproduced by the speaker in the radio when the "Talk" button is depressed.
3. In normal operation, acoustic feedback "howl" will occur when the AM radio receiver is tuned to a harmonic of the 200-kHz oscillator, and is placed near the intercom with the "Talk" button depressed.

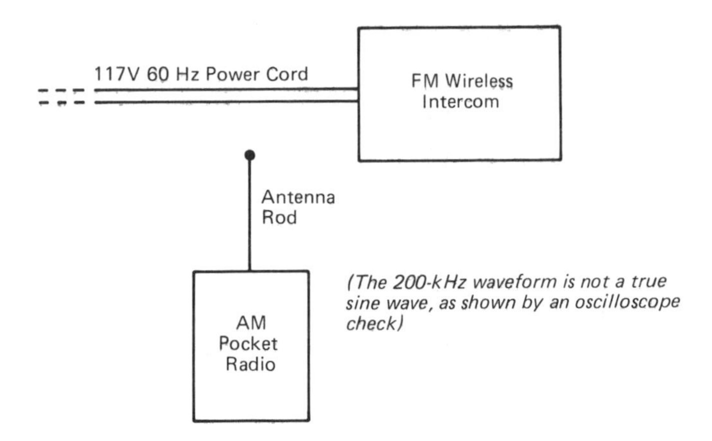

Note: Harmonics from the 200-kHz oscillator in the intercom will heterodyne with station carriers and will normally be audible at several places on the tuning dial of the AM radio receiver.

Quick Check: Tune the AM radio receiver to a harmonic of the 200-kHz oscillator, and turn up the volume control to a medium level. Then place the radio receiver near the intercom and depress the "talk" button. Normally, an acoustic feedback "howl" will result.

Figure 13-5 Quick check for intercom oscillator activity. The frequency-modulated signal has incidental amplitude modulation.

Case History—"No Reception, Transmits OK"

When the trouble symptom is "dead receiver," a systematic checkout is required. Power-supply trouble could be dismissed in this case, inasmuch as the transmitter section operated normally. DC voltage checks at transistor terminals in the receiver section were all within normal tolerance. Accordingly, the troubleshooter turned his attention to the input circuit (Figure 13-6).

An ohmmeter check showed that the switch was not defective. However, a resistance measurement across diodes D1 and D2 indicated zero ohms on the low-power function of the meter. (Normally, a reading of several ohms would be expected, due to the winding resistances of L1 and L2, plus the winding resistance of the input transformer primary.) Therefore, the troubleshooter concluded that there was a short-circuit somewhere along the input circuit.

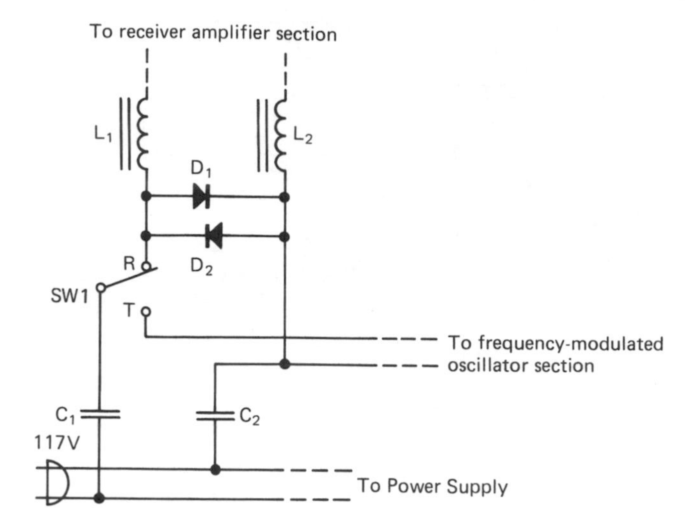

Note: Capacitors C1 and C2 couple the 200-kHz signal into and out of the 117-V line. When switch S1 is in the Receive (R) position, diodes D1 and D2 limit the signal amplitude that can enter the amplifier section via L1 and L2.

Figure 13-6 Input circuit to FM intercom unit. An oscilloscope is the most useful 200-kHz signal tracer.

A visual inspection did not turn up any evidence of a short between the PC conductors. Then, the troubleshooter disconnected diodes D1 and D2 for test; diode D1 measured zero resistance on the low-power ohms function of the meter, and D2 measured infinite resistance. When the shorted diode was replaced, the intercom was restored to normal operation.

The "dead receiver" trouble symptom described above was somewhat more difficult to tackle than for a wired intercom unit. If the audio amplifier is workable in a wired intercom unit, a hissing sound will be audible when the volume control is advanced and the speaker is held near the operator's ear. On the other hand, although the audio amplifier is workable in an FM wireless intercom unit, the speaker remains silent when the volume control is advanced.

TELEPHONE LISTENER

A telephone listener (portable phone listener) amplifies incoming calls so that others can listen in. It has loudspeaker output, and employs an inductive pickup with a suction cup that attaches or detaches easily from the receiver end of the handset. Volume control is provided; most telephone listeners are battery operated.

The circuit configuration consists of an inductive pickup unit, followed by a frequency-compensating RC feedback loop with a high-gain audio amplifier. When the operator advances the volume control and holds the speaker near his ear, a hissing sound is normally audible, mixed with a low-level 60-Hz hum due to stray-field pickup. However, if the inductive pickup unit is unplugged, the 60-Hz hum normally disappears, and the hissing sound remains.

Self-Oscillatory Quick Check

A self-oscillatory quick check for determining the workability of an amplifier in a telephone listener is depicted in Figure 13-7. The quick check consists of connecting a test lead from the speaker to the input of the amplifier. (The inductive pickup cable is removed.) When the volume control is turned to minimum, the telephone listener will normally produce a squealing sound from the speaker. As the volume control is turned up, the squealing sound will normally cease. In other words, the phase relations in the amplifier are such that positive

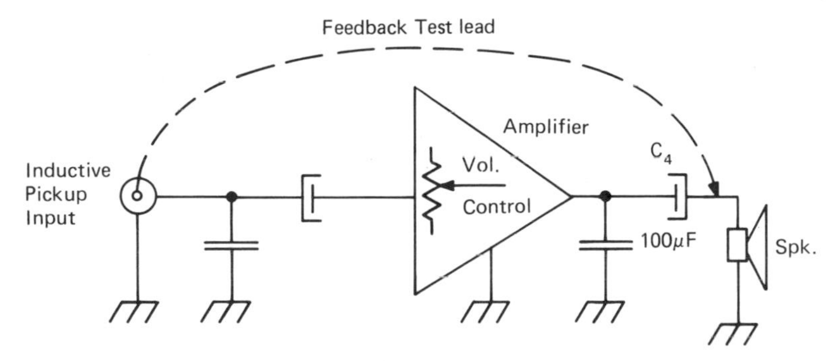

When troubleshooting a telephone listener without service data, and without a comparison unit, all of the quick checks explained in Chapters 1-3 will be found very helpful.

Note: This self-oscillatory quick check is suitable for the Archer 43-231A Telephone Listener. However, it is not necessarily suitable for other designs of telephone listeners. The troubleshooter should verify the applicability of any self-oscillatory quick-check arrangement on a known good telephone listener before concluding that it will be suitable.

Figure 13-7 Self-oscillatory quick check for telephone listener.

feedback occurs at minimum setting of the volume control, whereas negative feedback occurs at maximum setting of the volume control.

Case History—Weak Output

A telephone listener unit similar to the arrangement shown in Figure 13-7 developed a "weak output" trouble symptom. The battery voltage was normal, and a self-oscillatory quick check was made as a preliminary functional test. The unit failed the quick check, and there was no audible output from the speaker at any setting of the volume control. To make certain that the speaker was not defective, the troubleshooter checked for AC voltage across the speaker. There was zero AC voltage, but a DC voltage drop of 0.036 volt was unexpectedly measured.

Since the speaker is AC-coupled to the amplifier via C4 (Figure 13-7), there is normally zero DC voltage drop across the speaker. *Even a very small DC voltage across the speaker terminals is a definite indication of trouble in this type of circuit.* Therefore, the troubleshooter disconnected the capacitor and checked it for leakage. The capacitor was practically short-circuited, and measured 15 ohms leakage resistance. When the capacitor was replaced, the gain of the telephone listener unit was restored to normal.

ALTERNATIVE SELF-OSCILLATORY QUICK CHECK

With reference to the example of Figure 13-7, an alternative self-oscillatory quick check can be made by plugging in the pickup coil and turning up the volume control. A low-level 60-Hz hum is normally audible from the speaker, due to stray-field pickup. When the pickup coil is placed near the speaker, a motorboating sound or a squeal is normally audible. Whether the amplifier motorboats or squeals depends on the orientation of the pickup coil with respect to the speaker. If the telephone listener passes the quick check depicted in Figure 13-7, but fails a quick check when the pickup coil is plugged in, it is indicated that the pickup coil or its connecting cable is defective.

14

Closed Circuit Television Camera Troubleshooting

*Preliminary Trouble Analysis * CW Output, No Video * Pinpointing Defective Parts * Power Supply * Power Transformer Circuit * Horizontal Deflection * Vertical Deflection * Synchronizing Circuitry * Blanking Circuitry * Video Amplifier * VHF Modulator * Vidicon Section*

PRELIMINARY TROUBLE ANALYSIS

Closed-circuit television (CCTV) cameras use the basic sections depicted in Figure 14-1. A vidicon camera tube, deflection yoke, video amplifier, deflection circuits, sync waveshapers, and power supply are employed. The 1-inch vidicon camera tube forms the camera signal which in turn drives the video amplifier. This video amplifier supplies a complete picture signal at 1 V p-p into a 75-ohm load.

Note that the camera signal, sync pulses, and blanking pulses are combined in the video amplifier; sync action utilizes random interlace in this example. The video amplifier drives the VHF oscillator and modulator section which normally provides an output level of 50,000μV into a 300-ohm load on any channel from 2 to 5. Preliminary troubleshooting proceeds as follows:

CW Output, No Video. This is a comparatively common trouble symptom. It points to possible defects in the video amplifier, deflection section, or vidicon. *Case History:* The video signal was absent for 10 minutes after the CCTV camera was turned on; however, the VHF carrier signal was immediately outputted. The trouble was tracked down to a marginal electrolytic capacitor connected in series with the video-amplifier output.

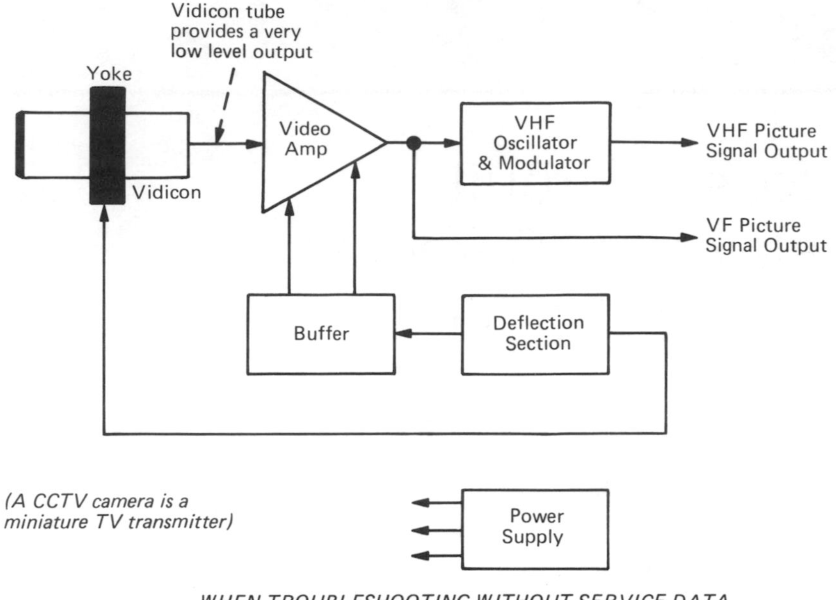

WHEN TROUBLESHOOTING WITHOUT SERVICE DATA, AND WITHOUT AN IDENTICAL CAMERA FOR COMPARISON TESTS, TRANSISTORS CAN BE "BUZZED OUT" AS EXPLAINED IN CHART 1-1. AN IN–CIRCUIT TRANSISTOR TESTER IS ALSO VERY HELPFUL IN PRELIMINARY TROUBLESHOOTING PROCEDURES.

Note: When there is no VHF picture signal output, and no video-frequency picture output, an audio-frequency test signal can be injected at the vidicon output to determine whether the trouble is in the camera circuitry or in the vidicon tube.

Troubleshooting without service data is greatly facilitated by the availability of an identical camera in normal operating condition for making comparative voltage, resistance, and impedance checks. If comparative checks are not feasible, the troubleshooter is thrown upon his own resources. In this situation, a comprehensive understanding of standard configurations and sectional functions is very helpful, so that the troubleshooter can meaningfully evaluate test measurements.

Figure 14-1 Block diagram of a typical CCTV camera.

Dead Camera. This is a less common trouble symptom. It points to power-supply failure, as detailed subsequently.

Video Output, No VHF Output. This trouble symptom points to a defect in the VHF oscillator and modulator section.

Camera Signal Weak. Check the vidicon terminal voltages; if ok, the tube should be checked by substitution.

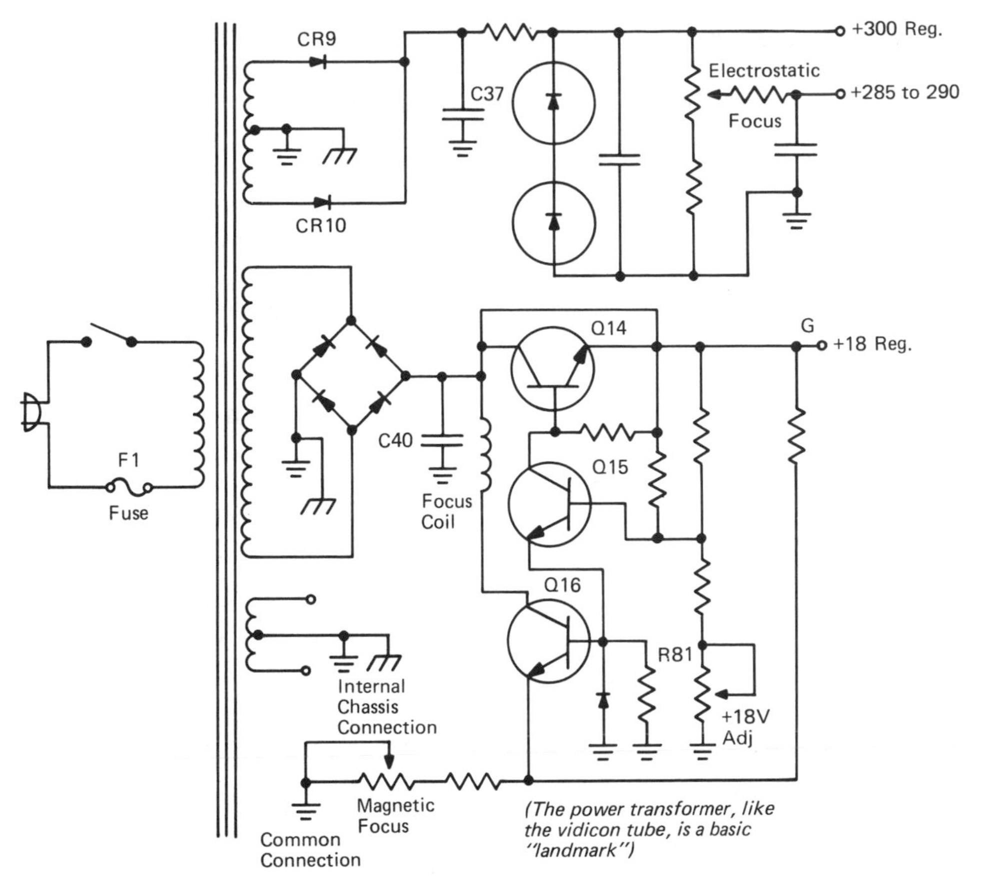

Note: The +18 V regulated output normally varies slightly with line-voltage fluctuation. For example, the following measurements are typical:

Line, 117.4 V, Reg, 18.06 V; Line, 110.1 V, Reg, 18.04 V; Line, 100.0 V, Reg, 18.03 V; Line, 95.0 V, Reg, 18.02 V; Line, 90.1 V, Reg, 17.96 V; Line 85.0 V, Reg, 17.50 V.

Figure 14-2 Power-supply circuitry for the exemplified TV camera. If feasible, comparative temperature checks can be helpful.

Pinpointing Defective Parts

From the troubleshooter's viewpoint, a CCTV camera consists of the following sections: power supply, focus-coil current regulator, vertical regulator, vertical deflection, horizontal deflection, blanking generator, video amplifier, and VHF oscillator-modulator sections.

Power Supply. Refer to Figure 14-2. First, a DC voltage

measurement should be made at point G. In this example, a reading of +18 volts is normally observed. An abnormal or a subnormal voltage value points to trouble in the power supply circuitry, or possibly to an excessive current demand by the camera circuitry.

In the case of power-supply trouble, the AC voltage at the power-line outlet should be measured. If the power-line voltage is normal, check next for a blown fuse (F1 in Figure 14-2). However, if the fuse is not blown, the next step is to remove the cover from the camera, plug in the line cord, turn on the power switch, and observe the vidicon tube to see whether its heater is glowing.

If the heater is dark, measure the heater supply voltage; this value is normally 6.3 V AC. If the heater supply voltage is correct, the heater is probably burned out; if the heater supply voltage is weak or zero, the power transformer is probably defective. *Note that even if the power-supply voltage is correct, operating trouble can be caused by excessive ripple*—measure the ripple voltage (if any) on the AC function of a DVM.

Power Transformer Circuit. To check the power-transformer circuitry, the CCTV camera's line cord is unplugged, and the resistance is measured between the prongs on the plug. With the power switch turned on, a reading of 65Ω is normally observed in this example. A reading of infinity can be caused by a *defective line cord*, a *defective switch*, a *blown fuse*, or a *burned-out primary winding*. A reading substantially less than 65Ω points to a partial short-circuit, such as short-circuited layers in the primary winding.

If the primary circuit checks out satisfactorily, the troubleshooter proceeds to measure the output voltages on the secondary side of the transformer. If one or more of the secondary voltages is subnormal, there is suspicion of short-circuited turns or layers in the secondary winding. (A zero reading usually results from an open-circuited winding.)

Next, if all of the secondary voltages measure correctly, the troubleshooter proceeds to make a DC voltage measurement at the positive terminal of C40 (Figure 14-2). A zero reading indicates that CR13 is probably defective. But if the measured value is normal (+26 V), the troubleshooter proceeds to point G and measures its voltage to ground (normally +18 V in this example). If this reading is zero, Q14 will probably be found defective; or if the reading is the same as at the positive terminal of C40, Q14 is probably short-circuited.

TABLE 14-1

Specified Device Terminal Voltages

Transistor	Emitter	Base	Collector
Q1	+ 3.7 V DC	+ 3.35 V DC	+15 V DC
Q2	0 (zero)	+ 0.55	+ 4.6
Q3	+ 4.0	+ 4.6	+ 8.8
Q4	0 (zero)	+ 0.65	+12.5
Q5	+ 0.75	+ 1.15	+11.0
Q6	+ 5.5	+ 6.2	+12.0
Q7	+11.5	+12.0	+18.0
Q8	0 (zero)	+ 0.3	+ 3.0
Q9	0 (zero)	+ 0.65	+ 1.2
Q10		(See Text)	
Q11		(See Text)	
Q12	0(zero)	− 1.8	+ 8.4
Q13	+ 0.4	+ 1.1	+ 7.3
Q14	+18	+18.5	+26
Q15	+ 6	+ 6.6	+18.5
Q16	+ 5.4	+ 6	+13

If the voltage at point G cannot be set to +18 volts by adjustment of R81, the remaining voltages in the regulator circuit should be measured. Specified DC voltages for this circuit are included in Table 14-1, for the exemplified circuitry.

Next, if the foregoing tests clear the regulated power supply from suspicion, the technician turns his attention to the 300-V supply in this example. Note that a DC voltmeter cannot show whether the voltage is pure DC or pulsating DC (DC with an AC ripple component). The voltage at the positive terminal of C37 in Figure 14-2 is normally 340 V. A zero reading indicates that CR9 and CR10 are probably defective; a short-circuit fault could also be present.

On the other hand, if all of the power-supply voltages measure specified value, the troubleshooter concludes that the malfunction is not in the power supply, and proceeds to the next logical probability.

Horizontal Deflection. An out-of-sync trouble symptom results if the oscillator in the horizontal-deflection section (Figure 14-3) runs too fast or too slowly to lock in on the associated TV receiver or video monitor. R56 is adjusted as required—however, if R56 is out of range,

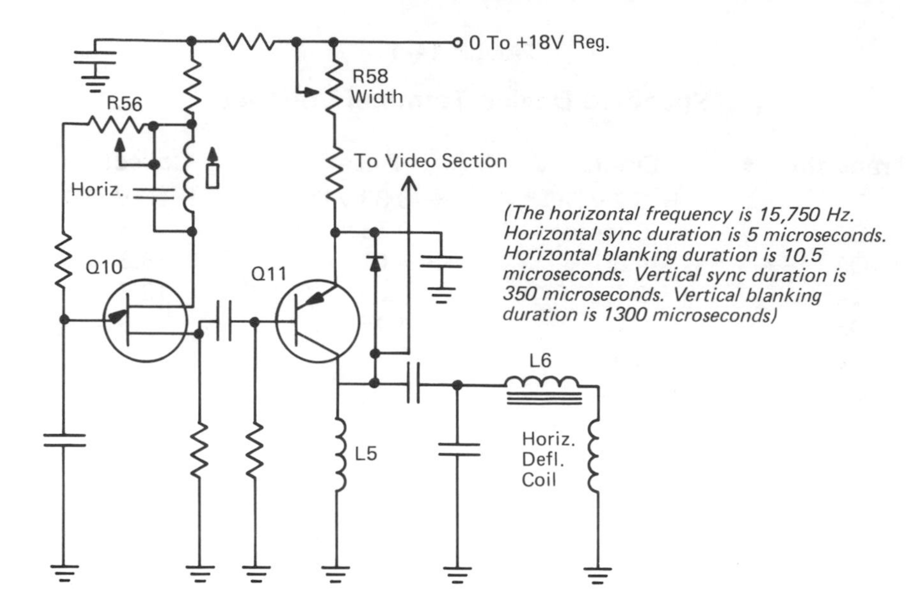

Caution: When troubleshooting the horizontal-deflection section, be careful not to underscan by setting R58 to a low resistance. This error could cause a raster burn.

Note: Specified voltage values (see text) are with reference to the +18 V regulated supply. Check the supply voltage and adjust to 18 V, if required, as explained under "Power Supply."

An oscilloscope is the most useful instrument for identifying the various circuit sections, provided that the troubleshooter recognizes the basic types of waveforms that are encountered in TV circuitry. (See also Chart 14-1.) When troubleshooting without an identical TV camera for comparison tests, the troubleshooter also needs to identify waveforms that have been subjected to various forms of distortion. With adequate knowledge of circuit action, an educated guess can be made concerning the circuit fault that is causing the waveform distortion.

Figure 14-3 Horizontal-deflection circuitry for the exemplified camera. If feasible, comparative temperature checks can be helpful.

Q10 will probably be found defective, in this example. If the pulse amplitude measures incorrectly, (normally, 65 V p-p, 10μs width, 15,750 Hz repetition rate), R58 is adjusted as required. However, if R58 is out of range, Q11 will probably be found defective. Normal DC voltages on Q10 and Q11 are:

Q10: emitter, +7 V; base 1, +0.6 V; base 2, +16 V.
Q11: emitter, +8.8 V; base, +8.8 V; collector, +4.2 V.

CHART 14-1

Differentiation and Integration of Basic Waveforms

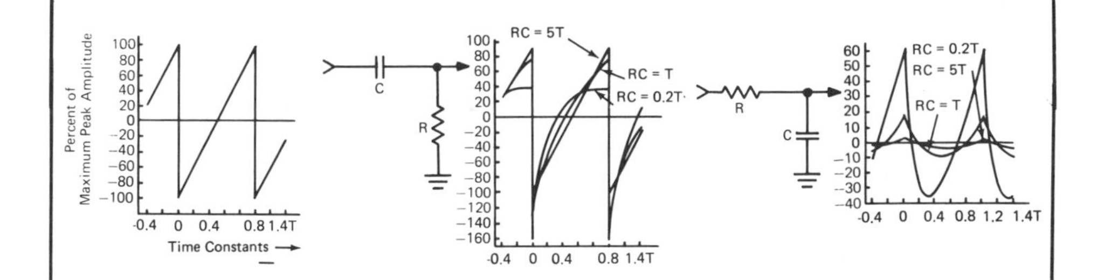

Sawtooth waveform with three degrees of differentiation and integration

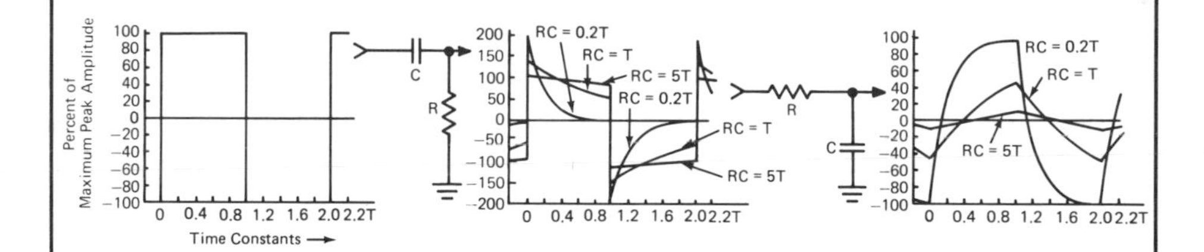

Square waveform with three degrees of differentiation and integration

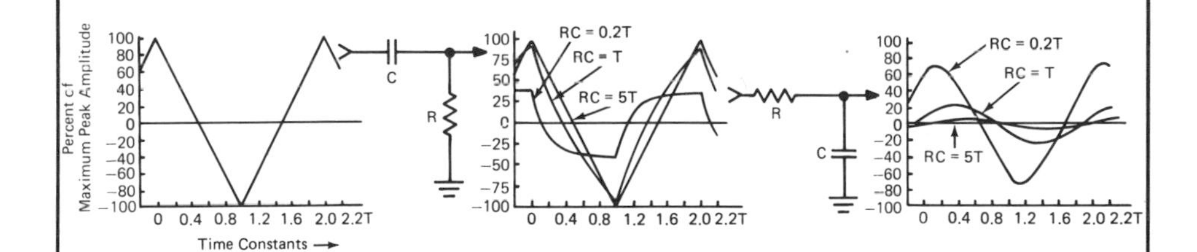

Triangular waveform with three degrees of differentiation and integration

Reproduced by special permission of Reston Publishing Co. and Robert Russell from Electronic Troubleshooting with the Oscilloscope.

A sawtooth waveform with an amplitude of approximately 7 V p-p is normally present at the emitter of Q10; it can be checked as previously explained. If this waveform is absent, Q10 is not oscillating. A positive pulse with a width of 10 μs and an amplitude of approximately 3 V p-p is normally present at base 2 of Q10. If this waveform is missing, Q10 is probably defective. The deflection yoke and coils L5 and L6 are checked by resistance measurements.

Vertical Deflection. An out-of-sync trouble symptom results if the oscillator in the vertical-deflection section runs too fast or too slow. An overall check of the vertical-deflection section exemplified in Figure 14-4 can be made by waveform measurement at the collector of Q13. Normally, a negative-going sawtooth with a 60-Hz repetition rate is present; it can be checked as previously explained. *If this waveform is weak, absent, or distorted, the troubleshooter proceeds to check the drive waveform into Q13.*

This is normally a positive sawtooth with an amplitude of approximately 0.5 peak-to-peak volt. Since its amplitude is comparatively low, a waveform check should be made with a preamp, as previously explained. The troubleshooter will also find it helpful to check the waveform at the collector of Q12. This is normally a negative-going pulse with an amplitude of approximately 7 V p-p, and a width of about 1.3 m.s. In case of zero output from Q12, check for a base driving waveform—if this waveform is missing, check for AC input at test points D and E, and at the primary of the blocking transformer T2.

To pinpoint a defective device or component, follow-up DC voltage measurements are made which may be supplemented by resistance measurements. Normal DC voltages for this example are listed in Table 14-1. Note that the vertical sweep amplitude is adjusted by setting R49 for an aspect ratio of 4 to 3 on the screen of the associated TV receiver or monitor. Vertical centering is adjusted by setting R55 to a point such that the DC voltages on either side of the vertical-deflection coil to ground are equal.

Synchronizing Circuitry

Out-of-sync trouble symptoms can also be caused by malfunctions in the synchronizing section. With reference to the example in Figure 14-5, the troubleshooter usually starts by measuring the DC voltage values at the transistor terminals. Specified voltages are listed in Table 14-1. Follow-up resistance measurements may be made with a low-power ohmmeter. If the troubleshooter needs

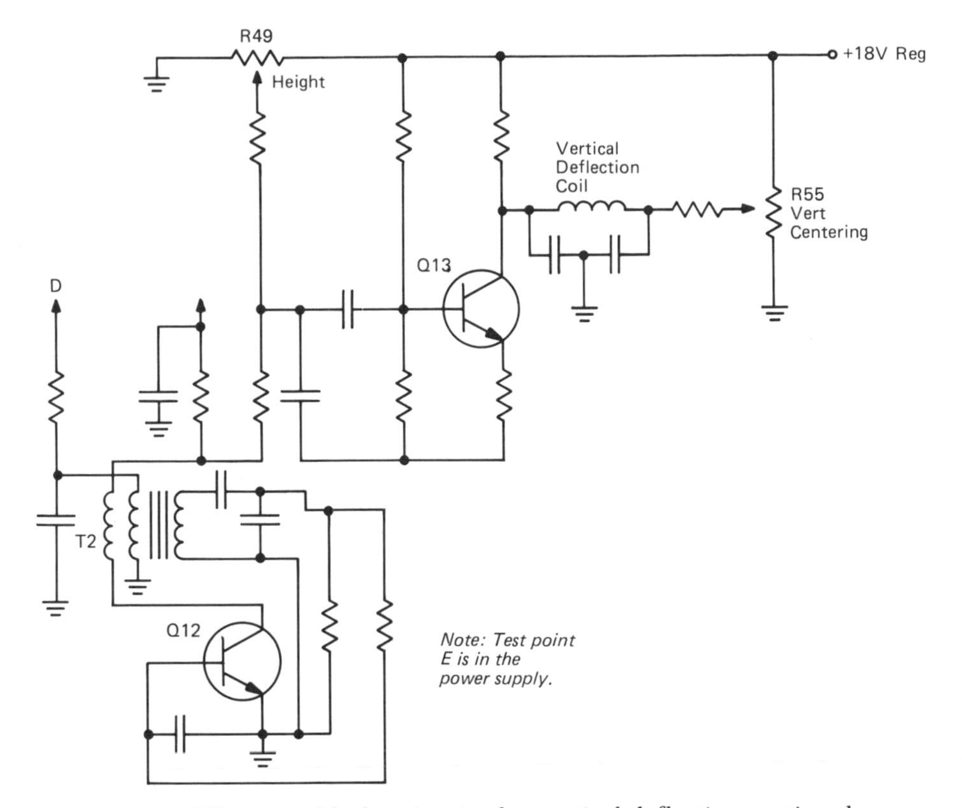

Caution: When troubleshooting in the vertical-deflection section, be careful not to underscan by setting R49 to a low resistance. This error could result in raster burn.

When comparative checks can be made on an identical camera which is in normal operating condition, temperature checks are very helpful, as explained in Chapter 1. Open capacitors, or capacitors with a poor power factor can be pinpointed with an impedance checker, as explained in Chapter 2.

Vertical-deflection circuitry can be spotted with an aural signal tracer, such as shown in Figure 2-1. The vertical-oscillator transformer is a helpful landmark.

Figure 14-4 Typical vertical-deflection circuitry. If feasible, comparative temperature checks can be helpful.

additional data, he or she may proceed by checking out the combined sync waveform at the collector of Q9. This is normally a mixture of horizontal sync pulses and vertical sync pulses. Their presence or absence can be verified by "beating out" 60-Hz and 15,750-Hz frequencies with an audio oscillator as previously explained.

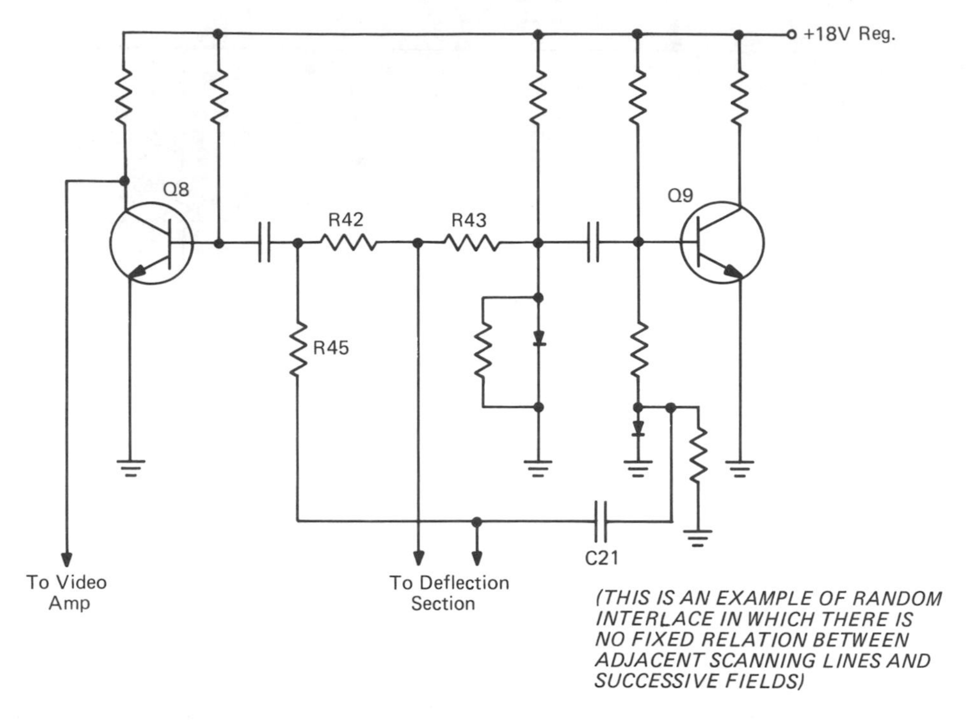

Note: The transistors can be quick-checked by means of a shut-off test. Apply a short circuit temporarily between the base and emitter terminals. In turn, the collector voltage will "jump up" to +18 V if the transistor is in normal condition.

To quick-check Q8, the collector output conductor to the video amp must be temporarily open-circuited—a razor slit may be made to open the conductor, and subsequently closed with a small drop of solder.

Figure 14-5 Synchronizing circuitry for the exemplified TV camera. The vertical frequency is power-line locked at 60 Hz.

1. If no 60-Hz component is found, the troubleshooter proceeds to check for the vertical-frequency input signal at the junction of R45 and C21.
2. On the other hand, if no 15,750-Hz component is found, the troubleshooter checks for the horizontal-frequency input signal at the junction of R42 and R43.

A check for the combined sync waveform can also be made at the base of Q9. This waveform normally has an amplitude of approximately 1.8 V p-p, in this example. Leaky capacitors are the most common culprits, followed by open capacitors. Caution: If a diode is replaced, be careful to observe correct polarity. *If a replacement diode is installed with reverse polarity, the troubleshooter is likely to have a tough-dog situation to contend with.*

Blanking Circuitry

Visible retrace lines in the reproduced picture point to a malfunction in the blanking circuitry. With reference to the example of Figure 14-5, the first step is to measure the DC voltage at the terminals of transistors Q8 and Q9. (See Table 14-1 for specified voltage values in this example.) Follow-up resistance measurements may be made with a low-power ohmmeter.

Then, if additional test data are needed, the troubleshooter should check out the combined blanking waveform at the collector of Q9, as explained above. If the vertical pulse (60-Hz fundamental component) is absent, the troubleshooter will check next for the vertical waveform at the junction of R45 and C21. On the other hand, if the vertical pulse (15,750-Hz component) is missing, the troubleshooter will check next for the horizontal waveform at the junction of R42 and R43. Another check for the combined waveform can be made at the base of Q8. Its normal amplitude is approximately 1.8 V p-p, in this example.

Video Amplifier

Weak, distorted, or absent output from the video amplifier can be investigated initially by bringing a finger near the case of Q3 (Figure 14-6). This is a quick check; if Q3 and the following stages are workable, considerable noise interference will be observed in the reproduced raster or picture. Troubleshooting proceeds by checking for sync signal at J1; the peak voltage is normally 1 volt on open circuit, or 0.4 volt into a 75-ohm load, in this example.

Defective components or devices can usually be pinpointed by means of DC voltage measurements. Specified video-amplifier voltages for this example are listed in Table 14-1. Follow-up resistance measurements with a low-power ohmmeter are often helpful. Note that the +18 V regulated supply must be set correctly, or measured voltage values will be deceptive. If the regulated voltage is off-value, set it correctly, as explained under "Power Supply."

Open capacitors are the most difficult fault to track down, because an open capacitor does not affect the normal DC voltage and resistance values. When a capacitor is suspected of being open, the troubleshooter usually bridges it temporarily with a known good capacitor to see whether normal operation is resumed.

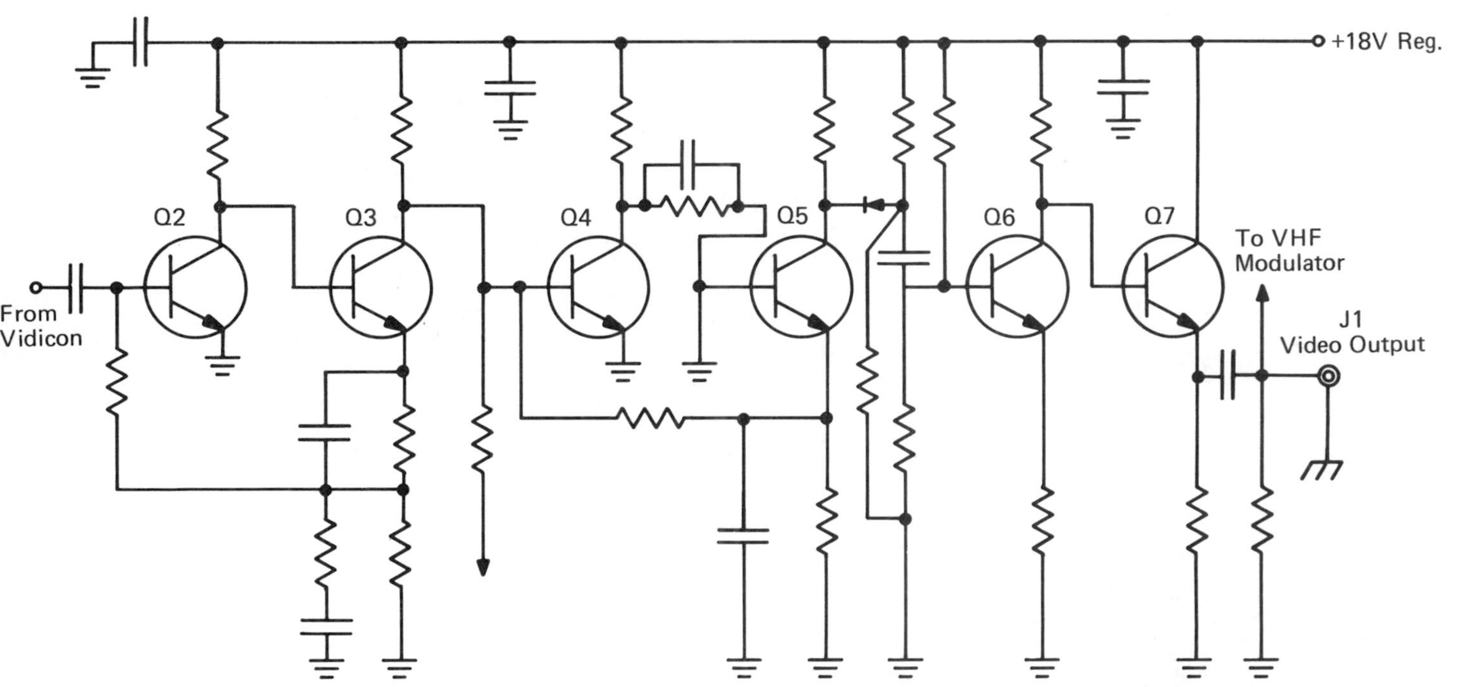

The video input transistor is easily spotted inasmuch as it is connected to the vidicon tube output. The video output transistor is also easily spotted inasmuch as it is connected to the video-output connector.

Note: The video amplifier has high voltage gain. An output of 1 V p-p is develped from a signal amplitude of 20 mV at the collector of Q2 (a gain of 50 times), and the signal level at the base of Q2 is not measurable with conventional service meters.

Caution: In this example, the TV camera employs a power transformer, and shock hazard is minimized. However, if you connect the video output connector to a TV receiver or a TV monitor that does not use a power transformer, a shock hazard will result accordingly.

Figure 14-6 Typical video amplifier circuitry. Rated frequency response is to 10 mHz.

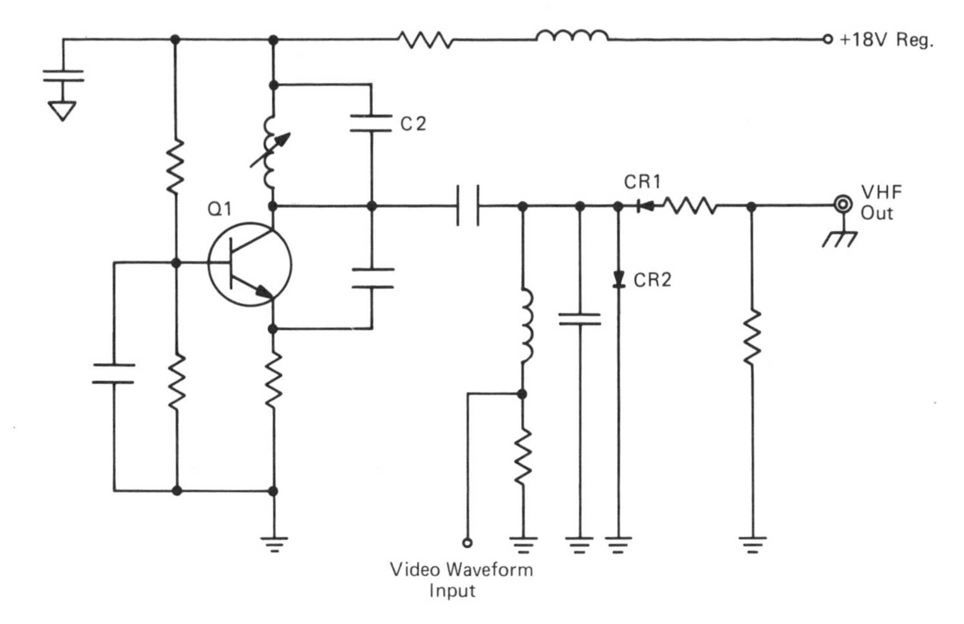

The VHF modulator diodes and VHF oscillator transistor are easily spotted because they are connected to the VHF output connector.

Note that if a bright light is placed behind the circuit board, the printed circuit conductors are readily visible through the board.

In this example, the rated modulated-RF output is 50 millivolts into a 300-ohm load. This output level is capable of driving 300 feet of coaxial cable.

Figure 14-7 Typical VHF modulator circuitry. The RF oscillator can be adjusted for operation on a chosen channel.

VHF Modulator

When there is normal video output, but weak or no modulated-VHF output, suspicion falls upon the VHF modulator section. (See Figure 14-7.) The troubleshooter starts by checking the video-input waveform to the modulator. This waveform normally has an amplitude of approximately 1 V p-p. If it is missing, attenuated, or distorted, diodes CR1 and CR2 should be checked next. To confirm an assumption that the oscillator is not operating, the troubleshooter makes a DC voltage measurement at the junction of CR1 and CR2.

If the oscillator is operating normally, a signal-developed bias of approximately −0.38 V will be measured. Oscillator malfunction is

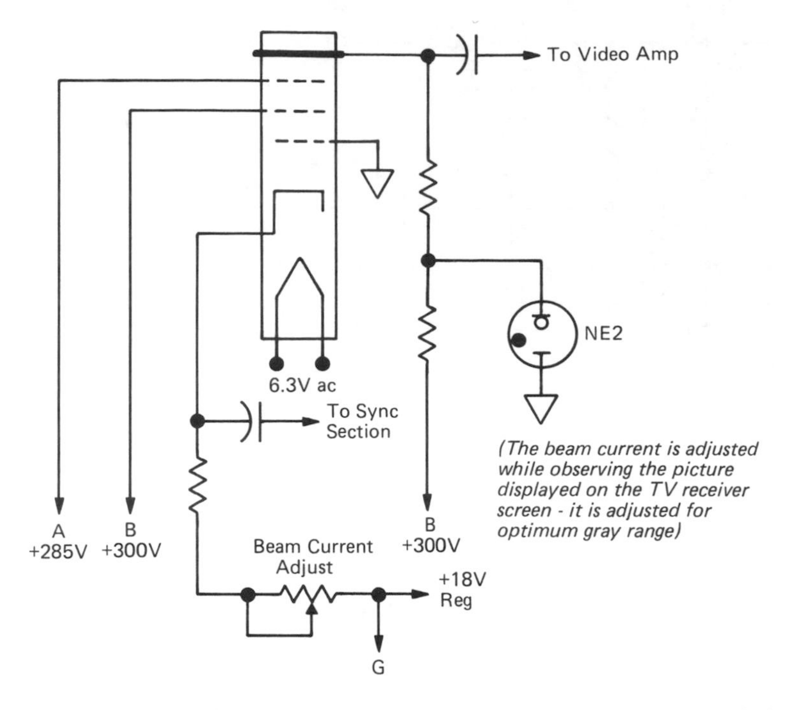

Caution: Avoid exposure of the vidicon screen to excessively bright light, such as direct sunlight—permanent screen damage can occur.

Figure 14-8 Typical vidicon circuitry. Electronic automatic beam current control is provided.

most likely to be caused by a defective transistor. Open capacitors can present puzzling malfunctions. For example, if C2 is open, the normal oscillating frequency jumps up, and the associated TV receiver will have a blank screen. Troubleshooters usually check suspected open capacitors by temporarily bridging them with known good capacitors.

Vidicon Section

The 1-inch vidicon tube shown in the example of Figure 14-8 cannot reproduce a good image unless the CCTV camera lens is carefully focused and the iris adjusted for optimum contrast under the prevailing lighting conditions. If too much light is admitted into the vidicon tube, the contrast will be abnormally high and picture detail will be degraded.

On the other hand, too little light into the vidicon results in subnormal contrast and noticeable noise interference in the image.

Note that a vidicon tube will be damaged immediately if it is directed toward the sun or any intense light source unless a suitable filter is used with a reduced iris opening. Note also that after a still scene has been scanned for some time, a noticeable negative after-image will persist for a while when the camera is directed at a blank wall, for example. However, this after-image will gradually decay with continued scanning action.

Troubleshooting of the vidicon circuitry starts with checking of the DC supply voltages. Observe whether the neon bulb NE2 is glowing—this gas tube regulates the target voltage in the vidicon. Follow-up resistance measurements may be made with a low-power ohmmeter. The coupling capacitors should be checked for leakage or open circuits before assuming that the vidicon tube is defective. If the vidicon is replaced, note that the focus adjustment of the camera is accomplished by repositioning the entire vidicon, yoke, and focus coil assembly.

15

Digital Troubleshooting Without Service Data

*Gates * Gate Resistances * Basic Gate Experiments * "Bad" Level * AND-Gate Arrangements * Digital Logic Probe * Piezo-Buzzer Comparator * Clock Subber * Clocked Counter Arrangement * Synchronous Ripple Counter*

GATES

Gates are the fundamental building blocks of digital circuits. A gate may stand alone, or it may be used in combination with other gates to form a flip-flop, counter, shift register, encoder, decoder, multiplexer, demultiplexer, or code converter. The four fundamental types of gates are called AND, OR, NAND, and NOR. Other forms of gates are termed XOR, XNOR, AND-OR, AND-OR-INVERT, and majority-logic gates. Although the troubleshooter is primarily concerned with TTL gates, he will occasionally encounter CMOS gates.

It is essential for the troubleshooter to have a clear understanding of gate action. If you are not familiar with gate functions, now is the time to reach for a basic text on digital logic, and to study gate operation. The troubleshooter requires more than theory—he or she also needs to know the various types of commercial digital integrated circuits with which he or she will be working.[1]

Gate Resistances

Although the checking of TTL gate resistances is not really new, many troubleshooters are not aware of this simple and useful aspect

[1]If you are unfamiliar with digital ICs, refer to pages 113 through 344 in *Encyclopedia of Integrated Circuits* by Walter H. Buchsbaum, Sc. D. (Prentice-Hall, 1980).

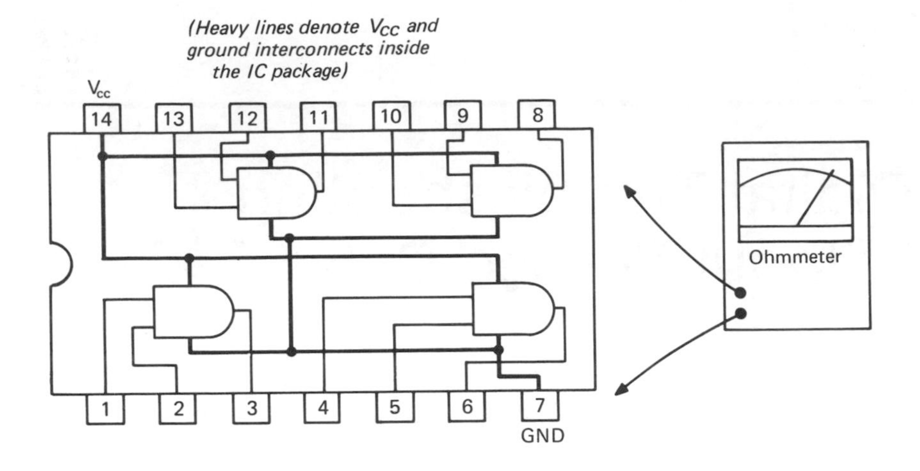

PIN 14 POSITIVE	
1. 16k	8. 28k
2. 16k	9. 16k
3. 28k	10. 16k
4. 16k	11. 28k
5. 16k	12. 16k
6. 28k	13. 16k
7. 9k	14. 0

PIN 14 NEGATIVE	
1. Inf	8. Inf
2. Inf	9. Inf
3. Inf	10. Inf
4. Inf	11. Inf
5. Inf	12. Inf
6. Inf	13. Inf
7. 3.3k	14. 0

PIN 7 POSITIVE	
1. 9k	8. 5.5k
2. 9k	9. 9k
3. 5.5k	10. 9k
4. 9k	11. 5.5k
5. 9k	12. 9k
6. 5.5k	13. 9k
7. 0	14. 3.9k

PIN 7 NEGATIVE	
1. Inf	8. Inf
2. Inf	9. Inf
3. Inf	10. Inf
4. Inf	11. Inf
5. Inf	12. Inf
6. Inf	13. Inf
7. 0	14. 9k

(These values are typical for a 50 kilohm/volt VOM on its Rx1k range. When a 20 kilohm/volt VOM is used on its Rx1k range, with pin 14 positive, typical readings are: 1.11k, 3.20k, and 6k.)

(The essential requirement is that readings be consistent, regardless of the ohmmeter used.)

Figure 15-1 Typical resistance values for a 7408 quad 2-input AND gate IC package.

of digital troubleshooting. Accordingly, typical resistance values for a 7408 quad 2-input AND-gate package are tabulated in Figure 15-1.

BASIC GATE EXPERIMENTS

The following experiments demonstrate some important characteristics of TTL gates, with which the troubleshooter should be familiar.

- Repeat some of the resistance measurements noted in Figure 15-1, and touch the tip of a soldering gun to the IC package. Observe the rapid change in resistance that results.
- Repeat some of the resistance measurements noted in Figure 15-1, using a 20,000 ohms-per-volt meter, and a 1,000 ohms-per-volt meter. Observe the indication differences between the two types of ohmmeters.
- Repeat some of the resistance measurements noted in Figure 15-1, using a 50,000 ohms-per-volt meter that employs a 9-V battery on its Rx10k range. Observe that zener breakdown of input junctions occurs on the Rx10k range.
- Make the input-output voltage experiment depicted in Figure 15-2 for an AND gate. Observe that electronic switching action is provided.

"BAD" LEVEL

Note the bad-region range in the foregoing experiment; this extends from 0.4 volt to 2.4 volts. The standard TTL logic-low range is from zero (ground) potential to 0.4 volt; the standard logic-high range is from 2.4 volts to 5.1 volts. If you measure 1.64 volts in a gate circuit, this value represents a bad level, and indicates that a defect is present.

AND GATE ARRANGEMENTS

Troubleshooters encounter various AND gate arrangements. For example:

- AND-gate action is sometimes provided by a NAND gate

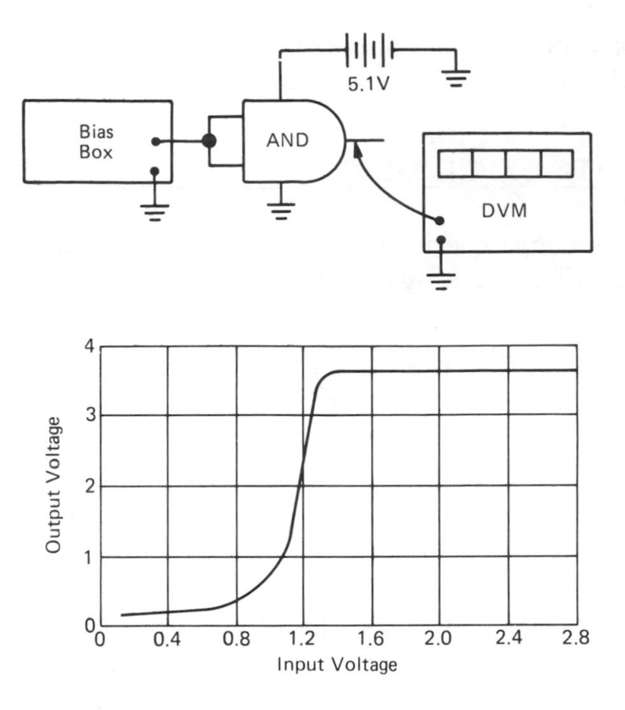

EXPERIMENT

This experiment shows the voltage transfer characteristic for an AND gate. In other words, it shows the relation of output voltage to input voltage.

A typical voltage transfer characteristic is shown in the plot. Observe that the output normally goes logic-high rapidly, although not instantaneously.

Stated otherwise, the switching characteristic of the AND gate traverses the "bad region" rapidly as the input voltage increases.

Both the logic-high and the logic-low levels of the gate are flat, because these levels correspond to saturated transistors in the gate circuitry.

Figure 15-2 Input-output voltage characteristic for an AND gate.

followed by an inverter, or by an inverter followed by a NAND gate.

- AND-gate action is sometimes provided by a NOR gate with inverters connected in series with its inputs.
- An AND gate is sometimes operated as a buffer; in this application, all of its inputs are tied together. (The logic

diagram often shows the AND gate with a single input, but when you buzz out the circuit, you will observe that it has more than one input terminal, all of which are tied together.)

- AND-gate action can be provided by an OR gate, with inverters connected in series with its input leads and output lead.

Similarly, more-or-less unexpected NAND-gate arrangements are likely to be encountered in troubleshooting procedures. For example:

- OR-gate action is occasionally provided by a NAND gate with inverters connected in series with its inputs.
- NOR-gate action may be provided by a NAND gate followed by an inverter, and with inverters connected in series with its inputs.
- Inverter action is commonly provided by a NAND gate with its inputs tied together.
- An important subclass of NAND gates is manufactured with open-collector output. These are used in wire-AND or wire-OR circuits. From the troubleshooter's point of view, these gates do not obey conventional NAND-gate truth tables. (Refer to a textbook on digital logic, if you are not knowledgeable about wire-AND and wire-OR.)

Again, somewhat unexpected XOR gate circuitry may be encountered. For example, you will find configurations in which an XOR gate has one input connected to V_{CC}. This arrangement operates as an inverter. Conversely, the XOR gate may have one input tied to ground; this arrangement operates as a buffer.

DIGITAL LOGIC PROBE

Almost everyone is familiar with digital logic probes—a logic probe is the most basic and most important digital tester. A logic probe is illustrated in Figure 15-3, with a logic pulser, a current tracer, and a logic clip. If you are in doubt concerning the troubleshooting applications of these four testers, refer to a digital troubleshooting book, or to the user's manuals for these testers. Otherwise, compounded confusion awaits in the IC jungle!

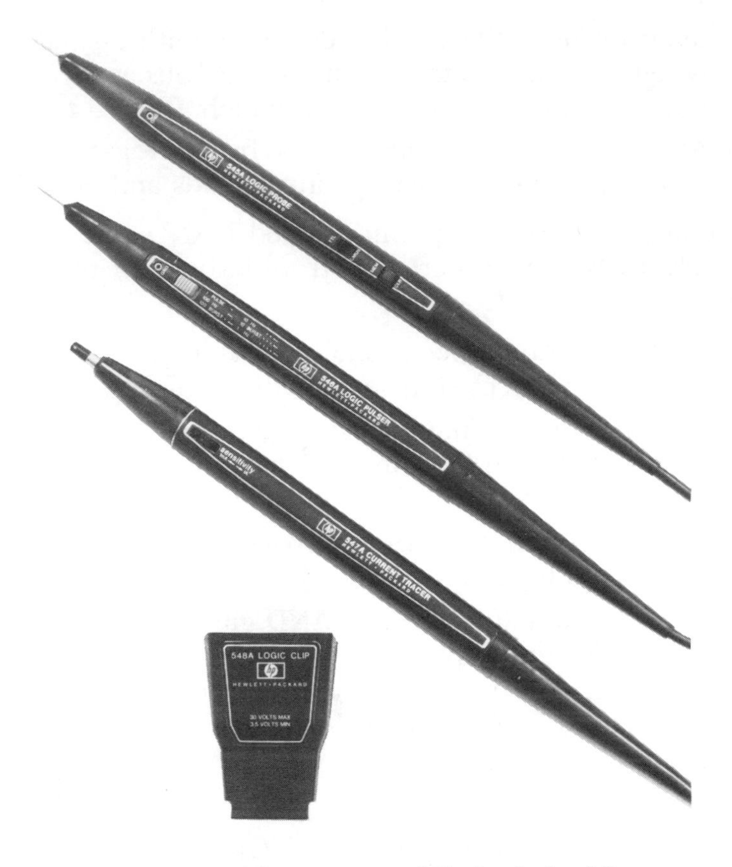

Photo, coutresy of Hewlett-Packard Company.

Figure 15-3 Logic probe, logic pulser, current/tracer, and logic clip.

LOGIC COMPARATOR

The logic comparator, illustrated in Figure 15-4, can be used for preliminary troubleshooting without regard to the associated circuit. In other words, the troubleshooter merely identifies the IC type number (marked on the IC package), and uses the logic comparator to check the IC against another known good IC of the same type.

Unless you are familiar with logic-comparator operation (and with the occasional limitations that you may encounter), now is the

Photo, courtesy of Hewlett-Packard Company.

Figure 15-4 Appearance of a logic comparator.

time to refer to a digital troubleshooting book, or to the user's manual for the logic comparator. Although a digital troubleshooter can get by without a logic comparator, it can lead to unnecessary difficulties. (The only exception is if you have a duplicate item of equipment in normal working condition, so that comparison tests can be made.)

PIEZO-BUZZER COMPARATOR

Most digital troubleshooters are familiar with piezo-buzzer testers for making logic-high and logic-low checks. A novel arrangement for making rapid comparison tests is depicted in Figure 15-5. The input leads A and B are applied at corresponding test points in the good unit and in the bad unit of digital equipment. If the piezo buzzer sounds, the troubleshooter knows that input leads A and B are not in

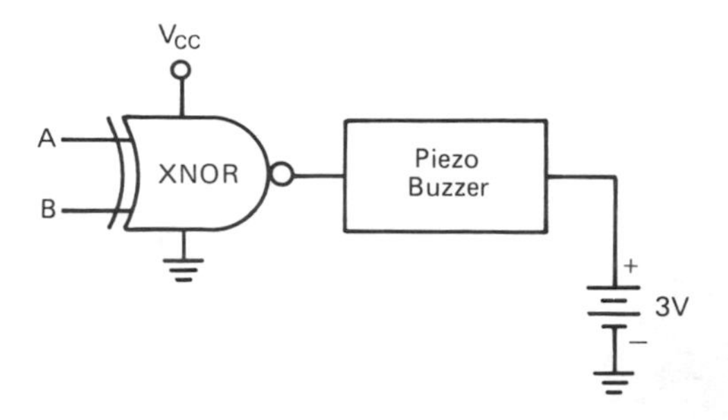

The XNOR gate has a logic-high output if inputs A and B are in the same logic state. On the other hand, the XNOR gate has a logic-low output if inputs A and B are in opposite logic states. As previously explained, the piezo buzzer will sound if it "sees" a logic-low source.

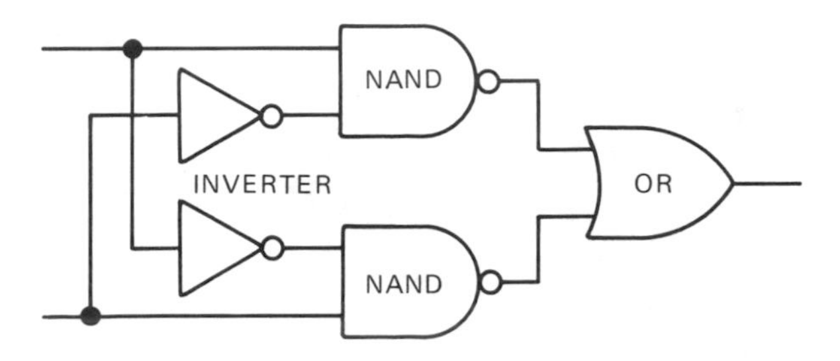

A common implementation of XNOR gate circuitry employs a pair of inverters, a pair of NAND gates, and an OR gate. The OR-gate output is logic-high unless the XNOR inputs are in opposite logic states.

Figure 15-5 A useful piezo-buzzer comparator that the troubleshooter can easily construct.

the same logic state. In turn, he or she starts looking for a fault associated with the test point in the bad unit.

A highly practical advantage of comparison checking is that the troubleshooter does not need to known how the digital circuit operates, nor whether the point under test is normally logic-high or logic-low. Of course, after a test point shows an incorrect state, the troubleshooter must then proceed to find out why. Sometimes the IC is faulted, and sometimes the node under test is faulted.

Comparison checks can be made with the digital equipment in a resting or operating condition. In either case, the inputs of the good unit and the bad unit must be maintained in the same states. Thus, if comparison checks are made with the equipment in a resting (static)

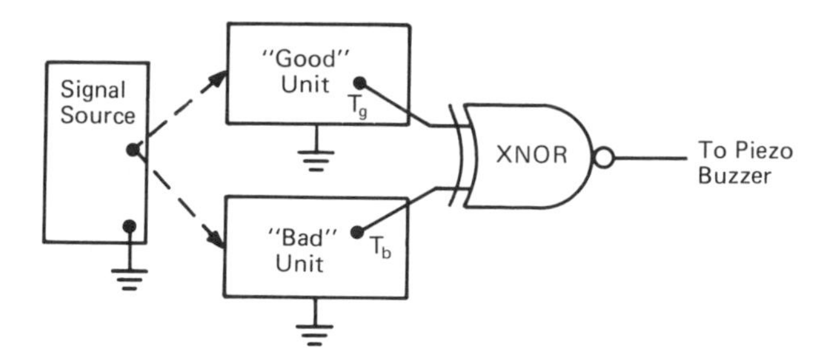

Note: Comparison tests with an XNOR comparator require that the same signal be applied to the inputs of the good unit and the bad unit. Otherwise, corresponding logic states would not be produced simultaneously at corresponding test points in the two units.

The corresponding test points in this example are T_g and T_b. A test point is usually a pin on an IC package. Thus, if T_g is pin 3 on a flip-flop IC in the good unit, T_b will be pin 3 on the corresponding flip-flop IC package in the bad unit.

Figure 15-6 Comparison tests require application of the same input signal to the good and bad units.

condition, both inputs may be connected to ground. Alternatively, both inputs may be connected to V_{CC}.

With reference to Figure 15-6, if comparison tests are made with the digital equipment in an operating (dynamic) condition, the same input signal must be simultaneously applied to the good unit and to the bad unit. Troubleshooters often use a square-wave generator as a signal source. The signal repetition rate should be quite slow in this test procedure, so that any sound output from the piezo buzzer will be clearly evident.

CLOCK SUBBER

Digital troubleshooters encounter both asynchronous and clocked logic circuitry. As asynchronous device has an operating speed that is not related to any frequency in the system to which it is connected. As an illustration, the asynchronous counters shown in Figure 15-7 have configurations wherein level changes occur at random times (the flip-flops are unclocked). Observe that when one flip-flop changes state, this state change triggers a second flip-flop (and so on).

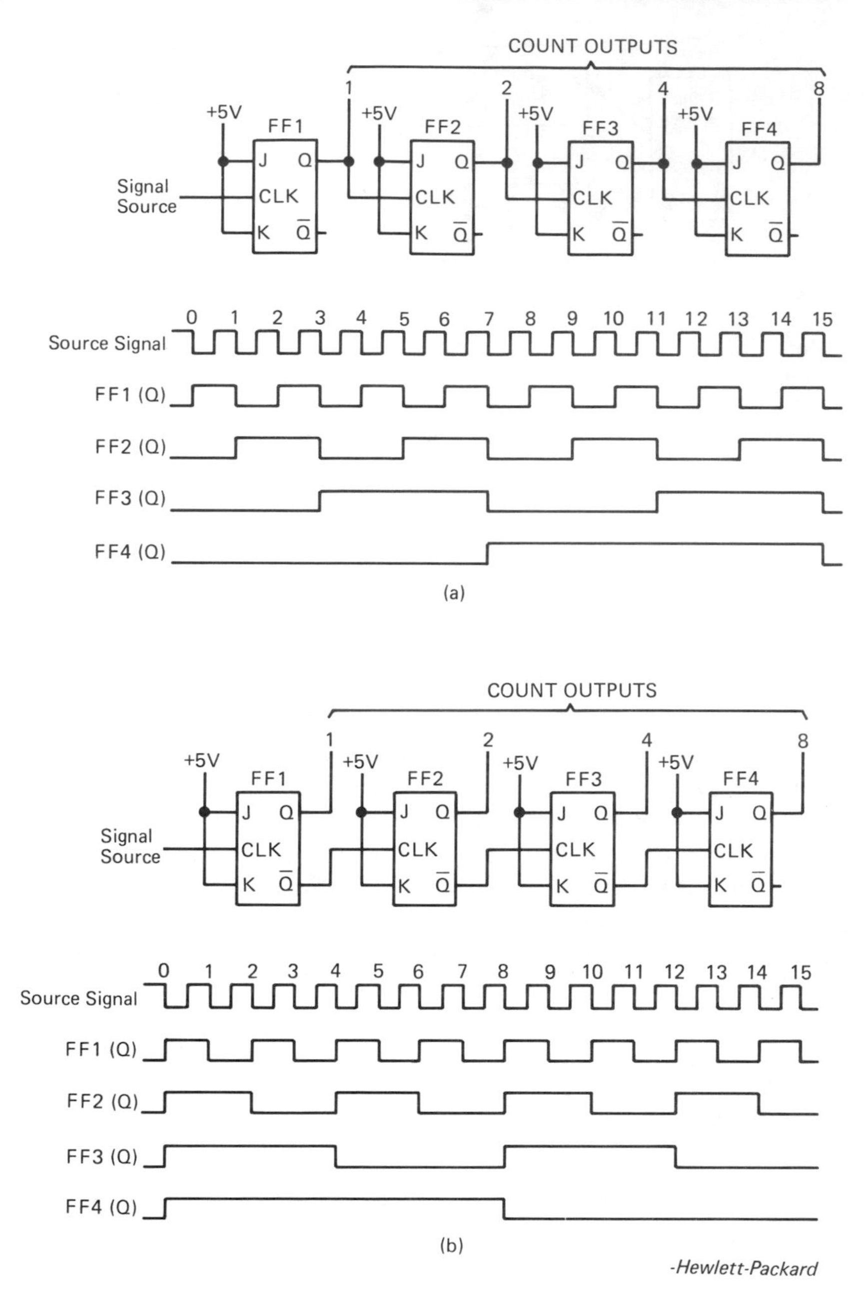

Figure 15-7 Counter circuits and timing diagrams. (a) Asynchronous count-up configuration; (b) asynchronous count-down configuration.

This counter configuration may have a misleading appearance to the apprentice technician, inasmuch as each flip-flop has a clock input terminal. Note, however, that each clock input is operated as a signal input in this asynchronous configuration. In other words, there is no clock-signal source, and the clock inputs are driven by signal sources. The clock inputs are employed as toggle inputs.

As a practical note, although the asynchronous counters in Figure 15-7 operate by a "domino" effect, the propagation times are quite precise. This is just another way of saying that each flip-flop may have a rated propagation time of 20 ns, for example. In normal operation, this value of time is precise—the tolerance on propagation time is tight. This means that if two units of equipment are driven from the same signal source, as depicted in Figure 15-6, the troubleshooter will expect that state changes throughout the configurations will occur simultaneously at corresponding test points (in normal operation).

Clocked Counter Arrangement

Next, observe the test setup depicted in Figure 15-8. Here, the bad and good units are clocked—they employ synchronous logic. A

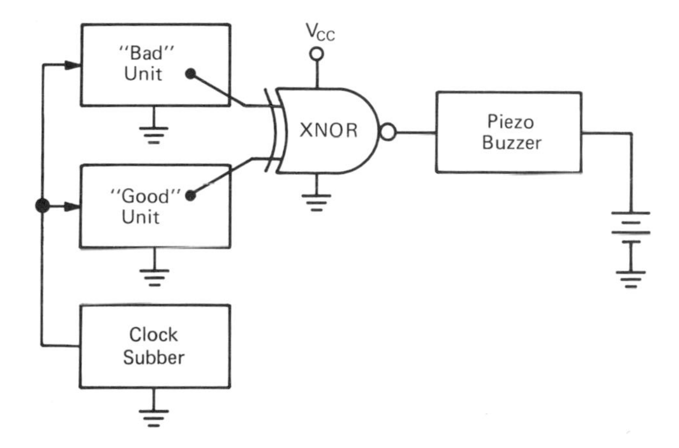

Note: In this example, the bad unit and the good unit are identical (except that the bad unit has a malfunction). Both units are clocked, and the test setup provides identical clocking for the units from a clock subber.

Figure 15-8 Digital equipment that employs synchronous logic is timed by a "clock."

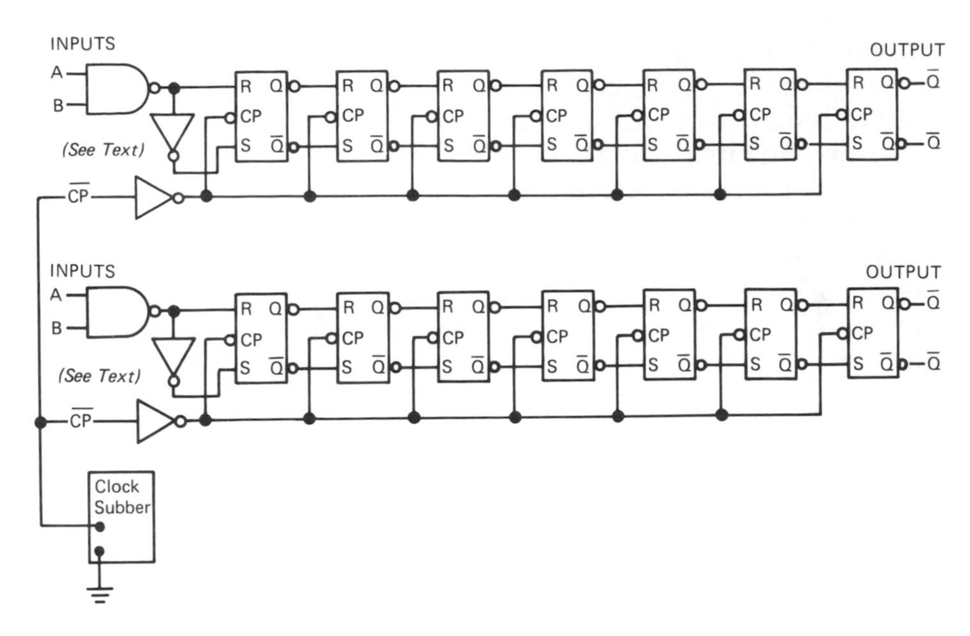

Technical Note: This is the simplest possible form of synchronous counter; it comprises flip-flops without any supplementary gates to speed up the carry process. In turn, the carries must ripple through the chain of flip-flops before the readouts (Q output states) are finalized. This process is called ripple-through carry.

Figure 15-9 All flip-flops in these two synchronous counters are triggered simultaneously by the clock-subber signal.

common clock signal triggers all of the flip-flops simultaneously, as seen in Figure 15-9. In other words, the clocks in the two counters are stopped, and the clock-subber signal is substituted for the system clocks. The clock subber is ordinarily a square-wave generator. As previously noted, the troubleshooter will usually wish to set the square-wave generator to a low repetition rate.

It follows from previous discussion that inputs A and B of both counters must have corresponding logic states in a comparison test. That is, both of the A inputs may be connected to V_{CC}, and both of the B inputs may be connected to ground, for example; or both of the B inputs may be connected to V_{CC}. In turn, the counters will either count up, or count down. Note the following terms:

1. In a sequential logic system, an output state is determined by the previous input state.

2. In a combinational logic system, the output state is determined by the present input state. (However, the troubleshooter will find combinational logic called sequential logic in some manuals.)
3. In synchronous operation, each event or operation is started by a clock signal.
4. In asynchronous operation, each event or operation is started by a signal which indicates that the previous operation has been completed.

Synchronous-Ripple Counter

Digital troubleshooters also encounter counter configurations which are neither synchronous nor asynchronous, in the strict sense of the terms. For example, the synchronous-ripple arrangement exemplified in Figure 15-10 is widely used. It is a simplified version of the true synchronous counter configuration which employs more elaborate gating circuitry.

Observe in Figure 15-10 that all of the flip-flops change state simultaneously in response to the clock signal. However, the

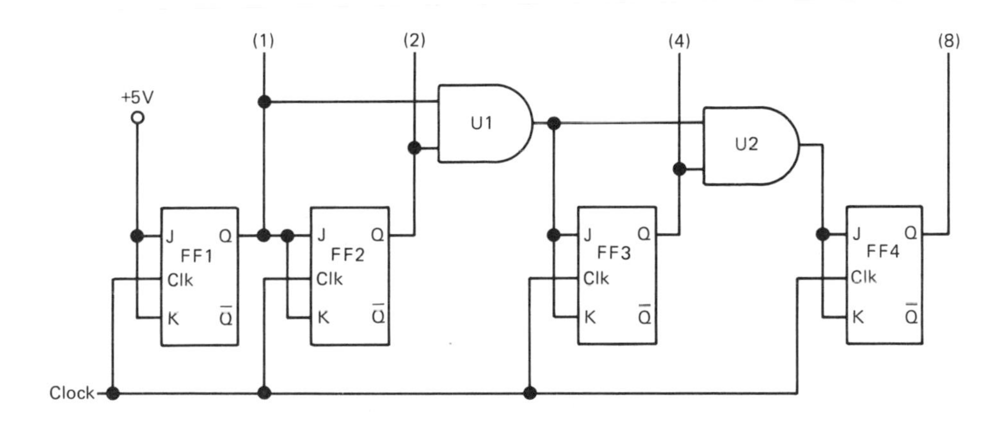

Note: When a high count capacity is provided in a binary counter, the synchronous-ripple arrangement such as that shown above can afford an attractive compromise between synchronous and asynchronous operating speeds, because its outputs change state simultaneously (ripple blanking is not required), and because of its minimization of hardware.

Figure 15-10 Example of a synchronous-ripple counter.

connection between the J and K inputs of a particular flip-flop and the Q outputs of all preceding flip-flops is made via AND gates that are connected in series. In turn, the propagation delay of each AND gate is cumulative, and the maximum counting speed is less than in a true synchronous counter configuration.

A true synchronous counter arrangement employs elaborated AND-gate circuitry whereby only one propagation delay is involved in the carry process, regardless of the number of flip-flops that are included in the chain. When a true synchronous counter is configured for high count capacity, many inputs must be provided for the AND gates, with many associated interconnections. In turn, production costs can determine the type of counter circuitry that the troubleshooter encounters in various types of digital equipment.

16

Progressive Digital Troubleshooting Without Service Data

*Glitch Grabber * Glitch-Grabber Application * Adjustable Single-Shot Pulser * NOR Gates in Wired-AND Operation * Feedback Experiment * AND Gates in Wired-AND Operation * Another Feedback Experiment * NAND Gate in Feedback Operating Mode * Digital System Noise * Generalized Digital Troubleshooting Procedure*

GLITCH GRABBER

A glitch is an unwanted false electronic pulse that masquerades as a legitimate digital pulse. Glitches can arise from various sources. For example, if a 2-input AND gate or NAND gate has a logic-high input and a logic-low input, the input signals can be complemented without changing the gate's output state (in normal operation). This complementation assumes that the input signals have very fast rise and fall, and that their corresponding leading and trailing edges occur simultaneously.

In practice, however, various malfunctions can result in less than ideal complementation. For example, consider an AND gate with its A input logic-high and its B input logic-low. The gate output is logic-low. Next, if the gate inputs are complemented, the A input is driven logic-low and the B input is simultaneously driven logic-high. If the rise and fall times of the driving waveforms are very rapid, the gate output remains logic-low.

On the other hand, if the drive waveform to the A input develops slowed fall, while the drive waveform to the B input maintains fast rise, there will be a brief moment during complementation when both inputs are logic-high. During this brief anomaly, the AND gate outputs a narrow pulse, or glitch. In turn, this unwanted false pulse is fed to the following device, where it is interpreted as a digital pulse.

Another source of glitches is timing skew, or an equivalent malfunction that causes complementary drive waveforms to arrive slightly earlier or slightly later than normally. As an example, if the A drive waveform were slightly delayed with respect to the B drive waveform, there would be a brief moment during complementation when both AND-gate inputs are logic-high. Accordingly, a glitch is outputted and is fed to the following device.

It should not be supposed that a digital system is free from glitches during normal operation. To the contrary, no device is ideal, and extremely narrow glitches are inevitably produced throughout a digital system. On the other hand, these "normal" glitches are so narrow that their energy content is very low, and they cannot trigger subsequent devices. Moreover, an extremely narrow glitch has very high frequency components; since digital circuitry has an upper cutoff frequency, it functions to attenuate extremely narrow glitches.

Marginal ground systems in digital circuitry can result in glitch production. Again, since digital circuitry is basically switching circuitry, small switching transients occur throughout a digital system. As exemplified in Figure 16-1, marginal circuit operation can lead to aggravated switching transients (glitches) that extend through the bad region and can cause false triggering of subsequent devices. Line-operated power supplies occasionally become faulty and introduce glitches into a digital system.

Troubleshooters often have considerable difficulty in tracking down glitch sources, and fall back on educated guesses and easter-egging approaches. Although a commercial logic probe will usually catch a glitch, an in-depth knowledge of circuit action plus close reasoning is usually required to identify the glitch source. This problem can be simplified in comparison test procedures, with the aid of a glitch grabber, as depicted in Figure 16-2.

GLITCH-GRABBER APPLICATION

Comparison tests facilitate glitch localization inasmuch as an opportunity is provided to screen out normal pulses, and to indicate the presence of a glitch. With reference to Figure 16-2, a simple glitch

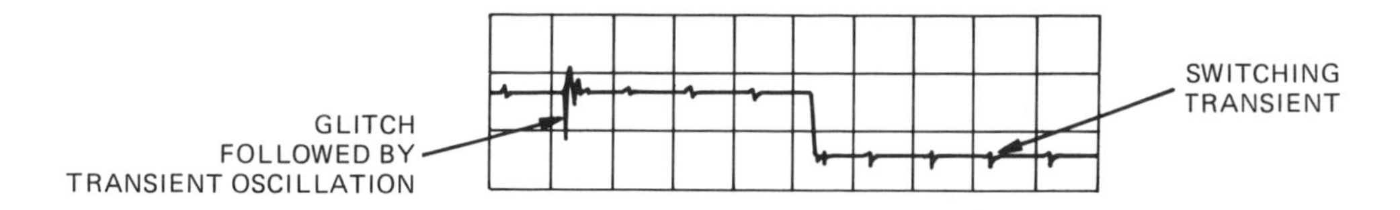

Courtesy of Hewlett-Packard Company

Figure 16-1 Example of a glitch and switching transients.

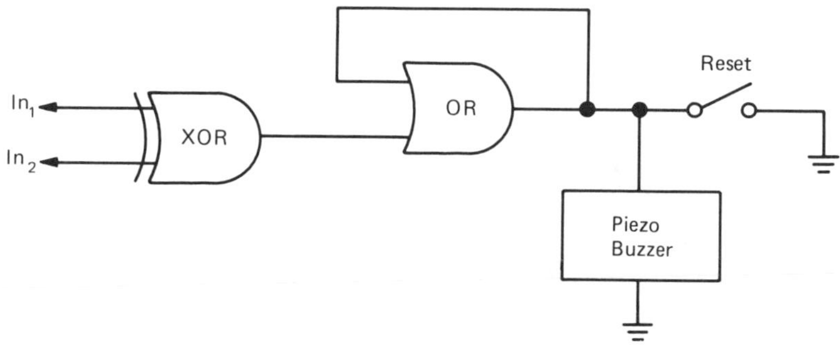

Figure 16-2 Arrangement of a simple glitch grabber.

grabber consists of an XOR gate, an OR gate, a piezo buzzer, and a pushbutton reset switch. The XOR gate functions to output a logic-high level if In_1 and In_2 are driven to complementary states. The OR gate functions as a glitch memory.

To understand how the OR gate functions as a glitch memory, observe that its output is fed back to one of its inputs. Consider first that the OR-gate output is logic-low, and that the XOR-gate output is logic-low. In this situation, both of the OR-gate inputs are logic-low, and its output is also logic-low; the piezo buzzer is silent.

Next, if the XOR-gate inputs In_1 and In_2 are driven to complementary states, the gate output goes logic-high. In turn, the lower input of the OR gate is driven logic-high, and the gate output goes logic-high. The piezo buzzer sounds off accordingly. Note that since the OR-gate output is fed back to one of the gate inputs, the gate is now locked logic-high. It is functioning as a glitch memory. The piezo buzzer will continue to sound off until the reset switch is momentarily closed. If the XOR-gate inputs are then in the same state, the piezo buzzer will be silent after the reset switch is opened. The glitch grabber is now armed, and waiting for the XOR gate inputs to be complemented.

Note that the glitch grabber screens out normal pulses in comparison tests because inputs In_1 and In_2 normally go logic-high or logic-low simultaneously. Accordingly, normal digital pulse activity does not result in a logic-high output from the XOR gate. However, if a glitch happens to occur in the bad unit, this glitch is not matched by a corresponding pulse in the good unit. In turn, the XOR-gate output goes logic-high for a brief moment, and the OR gate locks in on the event.

ADJUSTABLE SINGLE-SHOT PULSER

Marginally operating digital equipment was previously mentioned. When the troubleshooter is tracking down a marginal device, it may be helpful to determine the minimum voltage at which the device will trigger. Ordinary logic pulsers do not provide this facility. However, the adjustable single-shot pulser depicted in Figure 16-3 is easily constructed and serves the purpose satisfactorily.

In application, the capacitor charges up to a voltage which is determined by the setting of the 200-kilohm potentiometer. This voltage is indicated by a DVM. When the pulse switch is closed, the pulser outputs a surge with an amplitude equal to the DVM reading. Although the width of the surge pulse will vary greatly, depending upon the circuit resistance at the test point, the troubleshooter is concerned only with the surge amplitude.

NOR GATES IN WIRED-AND OPERATION

The troubleshooter may be confused if he happens upon NOR gates operating in wired-AND circuitry, as depicted in Figure 16-4. This is a seldom-used mode, although the 74 series of ICs may be configured in this manner. The essential point is to note that in the

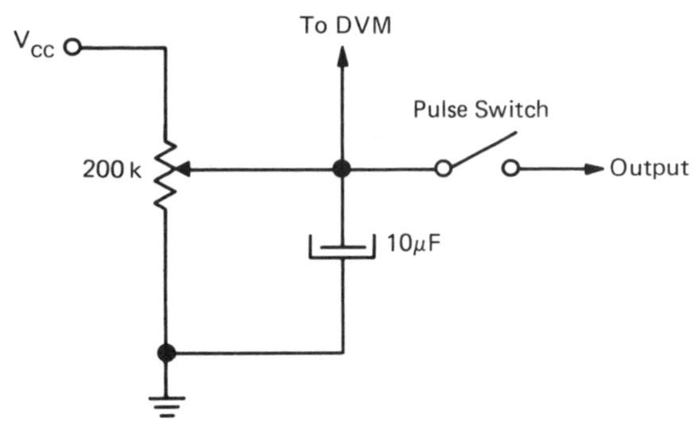

Figure 16-3 An adjustable single-shot pulser arrangement.

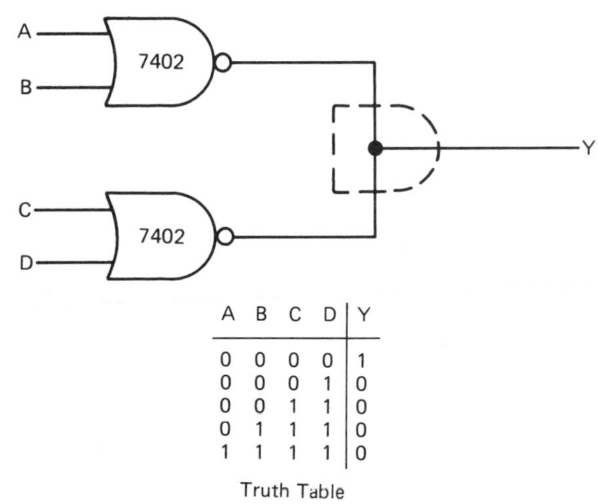

A	B	C	D	Y
0	0	0	0	1
0	0	0	1	0
0	0	1	1	0
0	1	1	1	0
1	1	1	1	0

Truth Table

Figure 16-4 NOR gates operating in wired-AND circuit.

wired-AND arrangement, if either NOR gate has a logic-high input, its output will go logic-low, and the output of the other NOR gate will thereby be restrained logic-low. Accordingly, the wired-AND circuit can normally produce a logic-high output only when all of the OR-gate inputs are simultaneously logic-low.

FEEDBACK EXPERIMENT

An informative feedback experiment with a NOR gate is shown in Figure 16-5. This configuration is illustrative of a short-circuit fault

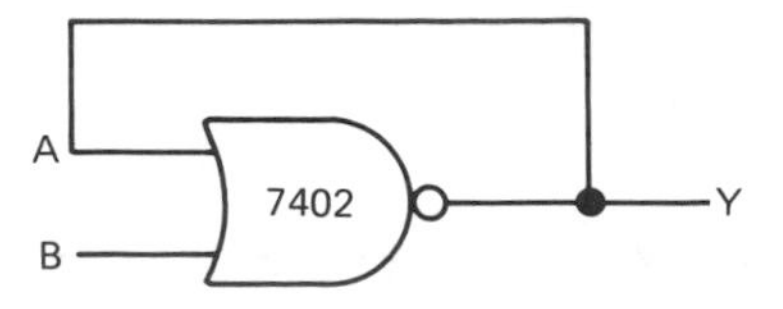

Note: When input B is grounded, output Y goes into the bad region at approximately 1.35 V.

When input B is connected to V_{CC}, output Y goes logic-low at approximately 0.08 V.

When input B is floating, Y goes logic-low at approximately 0.08 V.

Figure 16-5 NOR gate with its output fed back to one of the inputs.

in NOR-gate circuitry. Observe that the output from the NOR gate is fed back to one of the gate inputs. Circuit action is as follows:

1. When the B input is logic-high, the NOR-gate output goes logic-low.
2. When the B input is floating (open-circuited), the NOR-gate output goes logic-low. (The gate interprets the floating input as a logic-high level.)
3. When the B input is driven logic-low, the NOR-gate output goes into the bad-level region. (The gate output starts to go logic-high in response to the logic-low B-input level; on the other hand, a fed-back logic-high level starts to drive the output logic-low, and a necessary "compromise" level is established.)

Observe that a NOR gate with feedback does not function as a pulse memory, in contrast to OR-gate feedback operation. However, a negated-NAND gate with feedback functions as a pulse memory, in the same manner as an OR gate.

AND GATES IN WIRED-AND OPERATIONS

Wired-AND circuitry usually employs gates with open-collector output. An exception was noted in Figure 16-4, wherein NOR gates with totem-pole output were operated in a wired-AND configuration. When a short-circuit occurs between PC conductors in an AND-gate network, a pair of AND gates may thereby be forced to operate in the wired-AND mode, as depicted in Figure 16-6. This is a comparatively difficult troubleshooting situation; however, test-data analysis is

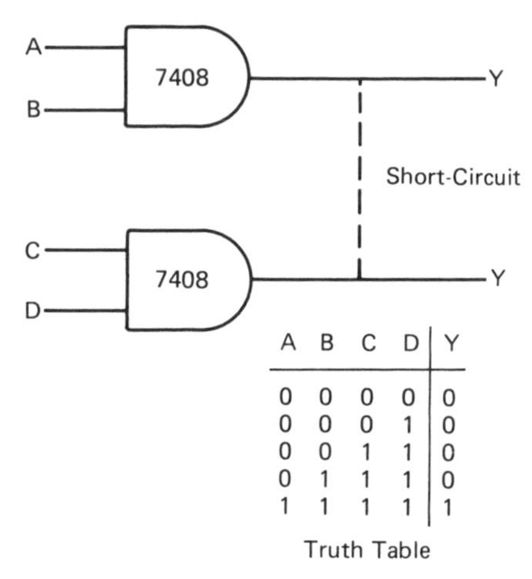

A	B	C	D	Y
0	0	0	0	0
0	0	0	1	0
0	0	1	1	0
0	1	1	1	0
1	1	1	1	1

Figure 16-6 An example of AND gates operating in the wired-AND mode.

facilitated by recognition of the abnormal circuit action that is involved.

The essential point to keep in mind is that since the AND gates in this example have totem-pole outputs, either gate may be driven to a logic-low output, and this logic-low state will constrain the other gate logic-low (the other gate cannot be driven logic-high). The bottom line is that both gates can be driven logic-low together, or both gates can be driven logic-high together. On the other hand, it is impossible to drive one gate logic-high if the other gate is being driven logic-low.

ANOTHER FEEDBACK EXPERIMENT

Another informative feedback experiment can be made with an AND gate, as depicted in Figure 16-7. Observe that the output from the AND gate is fed back to one of its inputs. The resulting circuit action is as follows:

1. When input B is grounded, the gate output goes logic-low and locks up logic-low. (If input B is floated, output Y remains logic-low.)
2. When inputs A and B are both driven logic-high, the gate output goes logic-high and locks-up logic-high. (If both inputs are floated, output Y remains logic-high.)
3. If input B is again grounded, the gate output goes logic-low and locks-up logic-low.

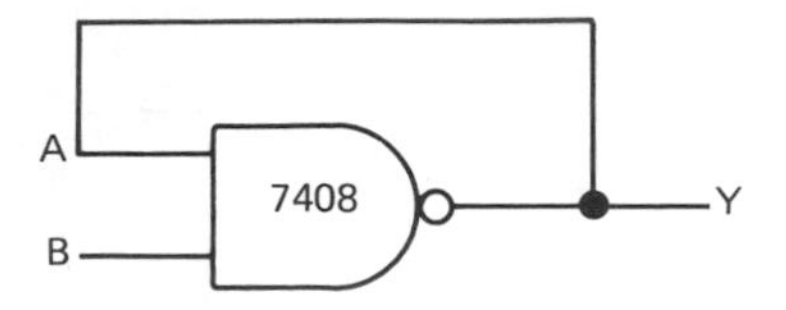

Note: When input B is floating (open-circuited), output Y may be locked-up either logic-high or logic-low. However, the lock-up condition can be broken from logic-high to logic-low by grounding input B.

Accordingly, the AND-gate feedback circuit does not operate as a pulse memory in the way that an OR-gate feedback circuit operates. (The OR-gate lock-up condition cannot be broken by driving the free input either logic-high or logic-low).

Figure 16-7 Experiment with AND gate operation in the feedback mode.

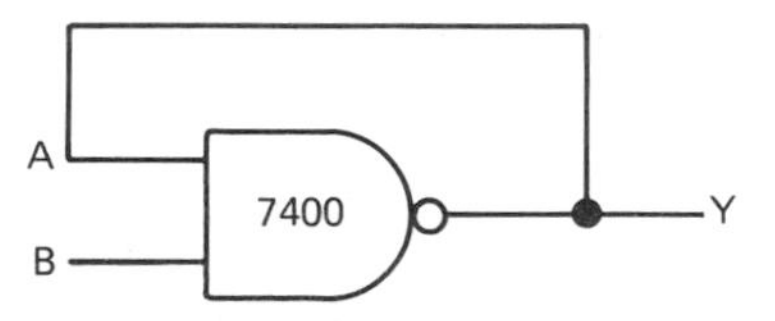

Note: Output Y can go logic-high, and this output state is reinforced when input B is grounded. However, when input B is floated (input B looks logic-high), a logic-high output state is not reinforced, but is opposed by feedback. In turn, Y assumes a bad-level potential.

The same circuit action prevails when input B is connected to V_{CC}.

Figure 16-8 Experiment with NAND gate operation in the feedback mode.

NAND GATE IN FEEDBACK OPERATING MODE

Consider next the circuit action that is obtained when a NAND gate is operated in the feedback mode, as shown in Figure 16-8. The logic states for three input conditions are as follows:

1. With input B grounded, the gate output Y goes logic high as a potential of approximately 4 V.
2. When input B is floated, Y assumes a bad-level potential of approximately 1.25 V.
3. With input B connected to V_{CC}, Y remains at a bad-level potential.

In comparison with the feedback action in the arrangement in

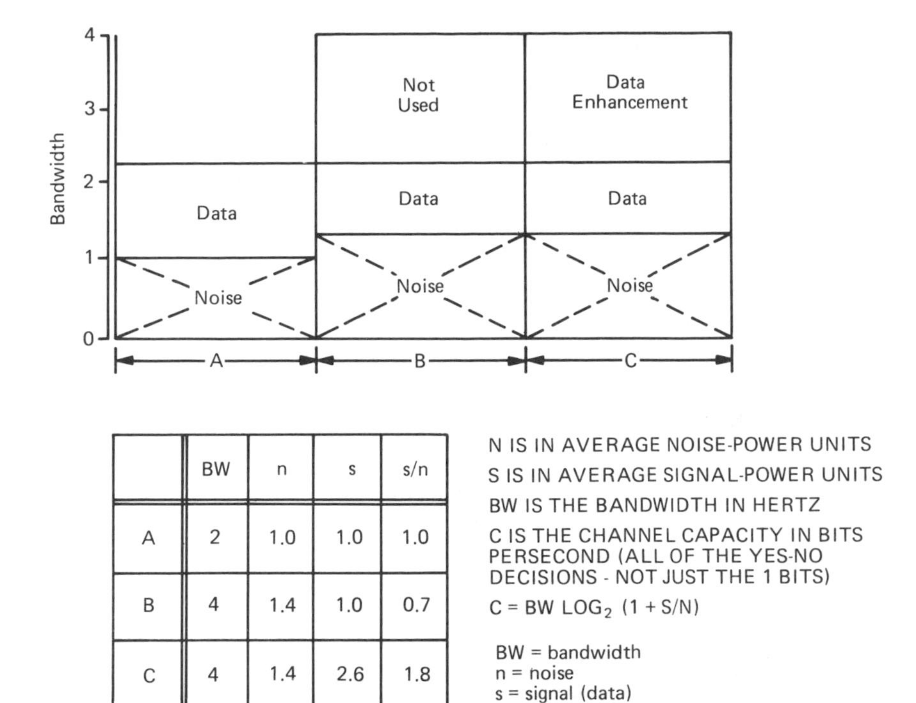

	BW	n	s	s/n
A	2	1.0	1.0	1.0
B	4	1.4	1.0	0.7
C	4	1.4	2.6	1.8

N IS IN AVERAGE NOISE-POWER UNITS
S IS IN AVERAGE SIGNAL-POWER UNITS
BW IS THE BANDWIDTH IN HERTZ
C IS THE CHANNEL CAPACITY IN BITS PERSECOND (ALL OF THE YES-NO DECISIONS - NOT JUST THE 1 BITS)

$C = BW\ LOG_2\ (1 + S/N)$

BW = bandwidth
n = noise
s = signal (data)
s/n = signal-to-noise ratio

Figure 16-9 Noise vs. bandwidth in a digital system.

Figure 16-7, lock-up is obtained when the AND-gate input is floating, because input B looks logic-high, and reinforces either a logic-high output or a logic-low output. On the other hand, lock-up is not obtained when the NAND-gate input is floating, because input B looks logic-high, and does not reinforce either a logic-high output or a logic-low output.

DIGITAL SYSTEM NOISE

Since there is always at least a trace of noise (AC noise) in a digital system, and because excessive noise can be a troublemaker, it is helpful to briefly recap the basic characteristics of AC noise. Any AC noise waveform has a rate of change that exceeds the response capability of the device under test. The AC noise immunity of a device denotes its capability of maintaining a given logic state with AC noise present.

The noise immunity of a device is measured with respect to noise-pulse width and the noise-pulse amplitude to which the device under test is unresponsive. Noise immunity is one of the basic measures of digital system reliability. If a device responds to a noise pulse, a data error is introduced. The meaning of noise versus bandwidth is depicted in Figure 16-9.

In A, the digital system has a bandwidth of 2 units (such as 2 kHz, or 2 MHz). In this example, the noise has a value of 1 unit, and the signal has a value of 1 unit, or the signal-to-noise ratio is 1. Next, in B and C, the bandwidth is twice as great; the noise increases to twice its value in A. The signal has the same value in B as in A (1 unit); accordingly, the signal-to-noise ratio decreases to 0.7.

Note that the remaining bandwidth capacity in B is not used. Observe in C that this remaining bandwidth capacity is used for data enhancement. In other words, the remaining bandwidth capacity is used to increase reliability by means of some form of redundant transmission. Data enhancement may employ actual redundant transmission (message repetition), special codes, multiple parity bits, or complement parity.

GENERALIZED DIGITAL TROUBLESHOOTING PROCEDURE

Although circumstances alter cases, it is helpful to note generalized digital troubleshooting procedures that take the majority of possible faults into account:

1. Use a logic comparator to check all of the permissible ICs in the suspected area. (Some digital ICs cannot be checked with a logic comparator.) Note down any ICs that fail the logic-comparator test.
2. Any ICs that cannot be checked with a logic comparator should now be tested with a logic probe, pulser, and clip.
3. A current tracer may then be used to track down faults in very low impedance paths, such as short-circuits to ground.
4. Note the occurrence of any pulse activity, but disregard the actual timing of the output signals. A logic probe is used to check for clock-pulse and input-signal activity, and if pulse activity is found on the output nodes, the troubleshooter assumes that the IC is operating normally.
5. The test data that have been noted in the foregoing steps are evaluated and analyzed to determine the cause of malfunction.

Index